Mathematical Devices for Optical Sciences

Mathematical Devices for Optical Sciences

Sibel Başkal
Department of Physics, Middle East Technical University, Ankara, Turkey

Young S Kim
Department of Physics, University of Maryland, College Park, Maryland, USA

Marilyn E Noz
Department of Radiology, New York University, New York, USA

IOP Publishing, Bristol, UK

ISBN   978-0-7503-1614-9 (ebook)
ISBN   978-0-7503-1612-5 (print)
ISBN   978-0-7503-1942-3 (myPrint)
ISBN   978-0-7503-1613-2 (mobi)

DOI   10.1088/2053-2563/aafe78

Version: 20190601

IOP Expanding Physics
ISSN 2053-2563 (online)
ISSN 2054-7315 (print)

British Library Cataloguing-in-Publication Data: A catalogue record for this book is available from the British Library.

Published by IOP Publishing, wholly owned by The Institute of Physics, London

IOP Publishing, Temple Circus, Temple Way, Bristol, BS1 6HG, UK

US Office: IOP Publishing, Inc., 190 North Independence Mall West, Suite 601, Philadelphia, PA 19106, USA

Contents

Preface

Galileo Galilei used his telescope before Newton wrote his books on mechanics and optics. More recently, in 1937, Francis Jenkins and Harvey White published their book entitled *Fundamentals of Optics*, which does not contain any matrices or make any use of Fourier transformations. This means that the science of optics was developed without matrices or Fourier transformations, yet we now know how essential these two mathematical devices are.

From the mathematical point of view, modern optics is the physics of two-by-two matrices and harmonic oscillators. The one- and two-photon coherent states are based on the mathematics of harmonic oscillators. The mathematical environment is the same for the information sciences which grew from the optical sciences.

In this book, we study the mathematical details of these two-by-two matrices. It is noted that the most general two-by-two matrix has six independent parameters if its determinant is to be 1, and is generated by six generators. The Lie algebra (closed set of commutation relations) for these six generators is the same as that of the Lorentz group which Einstein used as the mathematical basis for his special relativity.

There are many two-by-two matrices in the physics literature. Not all of them are generated by the three Pauli matrices. We thus need the six generators. In other words, we need the Lorentz group.

As for harmonic oscillators, there are many deformed and displaced Gaussian functions with one or two parameters in the optical sciences. The deformation takes the form of translation, rotation, and squeezing in phase space. It is a challenge to write the deformed Gaussian function as a series expansion of the complete orthonormal set of one-dimensional oscillator wave functions. It is noted that this problem is also an issue with the Lorentz group.

In this book, we start with the forms of quantum mechanics, namely the Schrödingner, Heisenberg, and interaction pictures. It is noted that the Wigner phase-space picture also plays an important role, in particular in addressing the issue of the Lorentz group.

We then deal with deformed and displaced Gaussian functions leading to coherent and squeezed states. We discuss the entangled two-parameter system in detail. It is shown that ignorance of one of the parameters leads to an increase of entropy in the measurable world.

As for optical instruments, it is noted that the basic reflections and refractions speak the language of two-by-two matrices, and thus the language of the Lorentz group. Lens optics, polarization optics, and cavity optics are discussed as illustrations.

The Lorentz group as presented in this book is based on the version originally developed for understanding another important aspect of physics. One hundred years ago, Bohr and Einstein met occasionally to talk about physics. Bohr was worried about the discrete energy levels of the hydrogen atom. Einstein was interested in how things look to moving observers. Did they ever wonder about how the hydrogen atom looks to moving observers? If they did, we do not know about it.

It was not possible then, nor is it possible today, to accelerate a hydrogen atom to relativistic speeds. However, although the proton is different from the hydrogen atom, it inherits the same quantum mechanics from the hydrogen atom. Therefore, we are now able to replace the bound state hydrogen atom by the proton which is a quantum bound state of quarks. Then to the moving observer it appears that the protons are coming from high-energy accelerators. This is not an issue in optical science, but is an issue where the fundamental physics of the Lorentz group plays the central role. We thus include in the appendix the Lorentz group applicable to Lorentz-boosted bound states.

SB, YSK, and MEN (November 2018)

IOP Publishing

Mathematical Devices for Optical Sciences

Sibel Başkal, Young S Kim, and Marilyn E Noz

Chapter 1

Forms of quantum mechanics

While there are several different but equivalent pictures of quantum mechanics [8], the Schrödinger picture is the one with which we are most familiar, as it is very convenient for atomic and nuclear physics. For quantum field theory and for making the correspondence with classical mechanics through the Poisson bracket, it is the Heisenberg picture that is the most useful. The interaction picture takes advantage of both the Heisenberg and Schrödinger pictures. It serves a useful purpose in time-dependent perturbation theory in both quantum mechanics and quantum field theory. Since these different pictures of quantum mechanics describe the same physics, choosing a particular picture of quantum mechanics over another depends on convenience.

In extracting measurable dynamical quantities from quantum mechanics, the concepts of the density matrix and probability distribution are needed. These are discussed in terms of mixed states, ensemble averages, and time dependence.

Indeed, quantum mechanics is built upon Heisenberg's commutation relations

$$[x_i, p_j] = i\delta_{jk}. \tag{1.1}$$

It is interesting to note the additional implications of this algebraic relation [1]. First of all, the commutation relations for the rotation group are direct consequences of Heisenberg's commutation relations. The rotation group can be extended to the Lorentz group which serves as the basic language for Einstein's special relativity.

In addition, Heisenberg's commutation relations can be written in terms of step-up and step-down operators applicable to the one-dimensional harmonic oscillator. This leads to the Fock space so essential for the present form of quantum field theory.

In section 1.1 the Schrödinger and Heisenberg pictures of quantum mechanics are introduced, and the interaction picture is illustrated in section 1.2. In section 1.3 quantum mechanics is formulated in terms of the density matrix. Section 1.4 contains further mathematical implications of Heisenberg's commutation relations, given in equation (1.1).

1.1 The Schrödinger and Heisenberg pictures

When the Schrödinger picture of quantum mechanics is used, only the wave functions describe the time evolution of the dynamical system. The dynamical operators themselves are time-independent. The situation is different in the Heisenberg picture where the time dependence is within the operators, and the state vectors are independent of time. We shall discuss here one example of the small number of problems which can be solved exactly in both pictures.

We consider a constant magnetic field in which a spin-1/2 particle is at rest. The time-dependent Schrödinger equation is

$$i\frac{\partial}{\partial t}\psi(t) = H\psi(t). \tag{1.2}$$

Here the operator and the wave functions are both defined in the Schrödinger picture. When the magnetic field is in the z-direction, this system has the Hamiltonian given by

$$H = \frac{1}{2}\begin{pmatrix} \omega & 0 \\ 0 & -\omega \end{pmatrix}. \tag{1.3}$$

With the normalization condition $|a|^2 + |b|^2 = 1$, the solution becomes

$$\psi(t) = \begin{pmatrix} a \ e^{-i\omega t/2} \\ b \ e^{i\omega t/2} \end{pmatrix}. \tag{1.4}$$

There is no change in the direction of spin when it is in the z-direction at $t = 0$. However, if the spin is initially along the x-direction with $a = b = 1/\sqrt{2}$ then the above solution takes the form

$$\psi(t) = \frac{\cos(\omega t/2)}{\sqrt{2}}\begin{pmatrix} 1 \\ 1 \end{pmatrix} - i\frac{\sin(\omega t/2)}{\sqrt{2}}\begin{pmatrix} 1 \\ -1 \end{pmatrix}. \tag{1.5}$$

Therefore, the probability that spin is up along the x-direction is $(\cos(\omega t/2))^2$ while the probability that it is down is $(\sin(\omega t/2))^2$. The above wave function can also be expanded in terms of spinors along the y-direction:

$$\psi(t) = \frac{e^{-i\pi/4} \ \cos(\omega t/2 + \pi/4)}{\sqrt{2}}\begin{pmatrix} 1 \\ i \end{pmatrix} + \frac{e^{i\pi/4} \ \sin(\omega t/2 + \pi/4)}{\sqrt{2}}\begin{pmatrix} 1 \\ -i \end{pmatrix}. \tag{1.6}$$

By taking into account these trigonometric relations

$$\cos^2(\omega t/2) = \frac{1 + \cos(\omega t)}{2}, \qquad \cos^2(\omega t/2 + \pi/4) = \frac{1 - \sin(\omega t)}{2},$$

$$\sin^2(\omega t/2) = \frac{1 - \cos(\omega t)}{2}, \qquad \sin^2(\omega t/2 + \pi/4) = \frac{1 + \sin(\omega t)}{2}, \tag{1.7}$$

we see that the probability of spin being along the y-direction is $\sin^2(\omega t/2 + \pi/4)$. Furthermore the spin precesses around the z-axis with the angular frequency of ω.

If we let σ_x and σ_y correspond to the spin operators along the x- and y-directions, respectively, the expectation value of σ_x and of σ_y become

$$\langle \sigma_x \rangle = \cos(\omega t) \quad \text{and} \quad \langle \sigma_y \rangle = \sin(\omega t). \tag{1.8}$$

Since, in the Schrödinger picture, the time dependence is strictly in the wave function, the operators do not depend on time. In contrast, in the Heisenberg picture the time dependence is in the operators. If we let σ_z correspond to the spin along the z-direction, the operator for the spin along the θ ϕ-direction is

$$A = \begin{pmatrix} Z & X - iY \\ X + iY & -Z \end{pmatrix}, \tag{1.9}$$

where

$$Z = \cos \theta, \quad X = (\sin \theta)\cos \phi, \quad \text{and} \quad Y = (\sin \theta)\sin \phi. \tag{1.10}$$

Considering that Heisenberg's equation of motion is

$$i\frac{d}{dt}(A_H(t)) = [A_H(t), H], \tag{1.11}$$

we find that X, Y, and Z satisfy the following differential equations:

$$\frac{dZ}{dt} = 0, \quad \frac{dX}{dt} - \omega Y = 0, \quad \text{and} \quad \frac{dY}{dt} + \omega X = 0. \tag{1.12}$$

This means that the direction of spin is now a function of time. Therefore, for the spin along the x-direction at $t = 0$, $Z = 0$, $X = \cos \omega t$, and $Y = -\sin \omega t$, and we have

$$A_H(t) = \begin{pmatrix} 0 & e^{i\omega t} \\ e^{-i\omega t} & 0 \end{pmatrix}. \tag{1.13}$$

Since this spinor corresponds to the spin precessing around the z-axis with angular frequency $-\omega$, we see that the rotation of this operator is in the opposite direction to the spin precession frequency obtained in the Schrödinger picture. The question now becomes: is this correct? To answer this we have to examine how the Heisenberg picture is related to the Schrödinger picture. It is possible to obtain the above spinor when we start from the Schrödinger equation. The differential equation's solution then becomes

$$\psi(t) = e^{-iHt}\psi(0). \tag{1.14}$$

Generally, a given time-independent operator A has an expectation value of

$$\langle A \rangle = (\psi(t), A(0)\psi(t)), \tag{1.15}$$

with $A(0) = A$. We can write this as

$$\langle A \rangle = (\psi(0), A_H(t)\psi(0)), \tag{1.16}$$

with

$$A_H(t) = e^{iHt} A(0) e^{-iHt}, \tag{1.17}$$

where e^{-iHt} is a unitary operator and carries out the time evolution from $t = 0$ to t. The wave function in the Heisenberg picture is $\psi(0)$, and is independent of time. Thus

$$\psi_H = e^{iHt} \psi(t). \tag{1.18}$$

Given the form of H in equation (1.3), we have

$$e^{iHt} = \begin{pmatrix} e^{i\omega t/2} & 0 \\ 0 & e^{-i\omega t/2} \end{pmatrix}. \tag{1.19}$$

In the Heisenberg picture the operator $A_H(t)$ should take the form of equation (1.13).

As we have seen above the Heisenberg operator is quite different from the time translation of the operator in the Schrödinger picture. The time evolution in the Schrödinger picture is given in equation (1.14), resulting in the time translation of a given operator as

$$A(t) = e^{-iHt} A(0) e^{iHt}. \tag{1.20}$$

This has a different form from that of $A_H(t)$ in the Heisenberg picture given in equation (1.17) because of the different sign of t. Since this operator satisfies

$$i\frac{\partial}{\partial t} A(t) = [H, A(t)], \tag{1.21}$$

it is quite different from Heisenberg's equation of motion expressed in equation (1.11). The above equation is often called Liouville's form of the Schrödinger equation. In this form,

$$A(t) = \begin{pmatrix} 0 & e^{-i\omega t} \\ e^{i\omega t} & 0 \end{pmatrix}. \tag{1.22}$$

Although this form of the Schrödinger equation is not widely known, it is used very frequently in atomic physics [2] and in quantum optics [10]. Because of the difference in the direction of time, the above operator is different from the time-dependent Heisenberg operator given in equation (1.13). In chapter 5 it is shown that Liouville's equation serves many useful purposes in the phase-space picture of quantum mechanics.

1.2 Interaction picture

It is possible to take advantage of the convenience of both the Schrödinger and Heisenberg pictures by using the interaction picture of quantum mechanics. In section 1.1, the transformation from the Schrödinger to the Heisenberg picture shown in the example could be easily performed because H is independent and the transformation matrix e^{iHt} takes a simple form. If, however, a complicated term is added to this simple Hamiltonian so that the total Hamiltonian is $H + G$, we no longer have a simple transformation. Nevertheless, with the simple portion of the total Hamiltonian, we can make a transformation.

To illustrate this point, we start with the example discussed in section 1.1 and add a weak sinusoidal magnetic field along the x-direction. The total Hamiltonian is then $H + G$, with H given in equation (1.3), and G given by

$$G(t) = b\begin{pmatrix} 0 & \cos \alpha t \\ \cos \alpha t & 0 \end{pmatrix}. \tag{1.23}$$

In the above equation the strength of the interaction is measured by the coefficient b. The Schrödinger equation then becomes

$$i\frac{\partial}{\partial t}\psi(t) = (H + G)\psi(t), \tag{1.24}$$

where H is a time-independent Hamiltonian. Let us consider the wave function in the interaction picture where it is defined as

$$\psi_I(t) = e^{iHt}\psi_S(t) \quad \text{or} \quad \psi_S(t) = e^{-iHt}\psi_I(t). \tag{1.25}$$

Then with

$$G_I(t) = e^{iHt}G(t)e^{-iHt}, \tag{1.26}$$

$\psi_I(t)$ satisfies the equation

$$i\frac{\partial}{\partial t}\psi_I(t) = G_I(t)\psi(t). \tag{1.27}$$

Here the operator $G_I(t)$ and the wave function $\psi_I(t)$ are called the Hamiltonian and the wave function in the interaction picture. It is clear that the differential equation of (1.27) is simpler than that of (1.24), particularly when b of equation (1.23) is small. This form of quantum mechanics is, indeed, the starting point for relativistic quantum field theory and quantum electrodynamics leading to Feynman diagrams [7]. In quantum optics, where the radiation and absorption of a photon is the main subject, this picture of quantum mechanics is important.

The essence of this picture can be understood by going back to the problem of a spinning particle in the magnetic field. From equations (1.19) and (1.23), $G_I(t)$ can be written as

$$G_I(t) = b(\cos \alpha t)e^{iHt}\sigma_x e^{-iHt}$$
$$= b\begin{pmatrix} 0 & e^{i\omega t}\cos \alpha t \\ e^{-i\omega t}\cos \alpha t & 0 \end{pmatrix}. \tag{1.28}$$

The fact that $G_I(t_1)$ and $G_I(t_2)$ do not commute with each other for different values of t_1 and t_2 presents the major difficulty in solving the Schrödinger equation of (1.27). Time ordering is therefore needed when we solve the equation by iteration. The solution is given by

$$\psi_I(t) = \left\{ \sum_{n=1}^{\infty} ((-i)^n/n!) P \int_0^t \cdots \int_0^t G_I(t_1) \cdots G_I(t_n) dt_1 \cdots dt_n \right\} \psi_I(0), \qquad (1.29)$$

where the time-ordering operator P dictates that $G_I(t_i)G_I(t_j)$ in the integrand be ordered so that $t_i > t_j$. For sufficiently small values of b, it is valid to take a few lowest order terms in the above series. If b is large, however, calculating measurable numbers from this series is not useful [11].

The rotating wave approximation is one of many other approaches to this problem. Let us write

$$\psi_I(t) = \begin{pmatrix} C_1(t) \\ C_2(t) \end{pmatrix}, \qquad (1.30)$$

then equation (1.27) can be written as [9]

$$\begin{aligned} \dot{C}_1(t) &= -ib\big(e^{i\omega t}\cos(\alpha t)\big)C_2(t), \\ \dot{C}_2(t) &= -ib\big(e^{-i\omega t}\cos(\alpha t)\big)C_1(t). \end{aligned} \qquad (1.31)$$

Although these equations appear to be very simple, they cannot be solved analytically. Currently, these equations can be solved numerically. This fact should not prevent us from studying the properties of these equations. It is easy to prove that the total probability is conserved:

$$|C_1(t)|^2 + |C_2(t)|^2 = 1. \qquad (1.32)$$

We note next that the difference between the two frequencies $(\omega - \alpha)$ can be much smaller than the sum $(\omega + \alpha)$. If the primary interest is in the frequency region where $(\omega - \alpha)$ is small, we can ignore the high frequency component of $(\omega + \alpha)$. Hence $G_I(t)$ can be written as

$$G_I(t) = \frac{b}{2} \begin{pmatrix} 0 & e^{i(\omega-\alpha)t} \\ e^{-i(\omega-\alpha)t} & 0 \end{pmatrix}. \qquad (1.33)$$

Then the differential equations in equation (1.31) assume the form

$$\begin{aligned} \dot{C}_1(t) &= -i\frac{b}{2}e^{i(\omega-\alpha)t}C_2(t), \\ \dot{C}_2(t) &= -i\frac{b}{2}e^{-i(\omega-\alpha)t}C_1(t). \end{aligned} \qquad (1.34)$$

The above differential equations can be decoupled if we impose the initial condition that $C_1(0) = 0$ and $|C_2(0)| = 1$. The solutions are then

$$\begin{aligned} C_1(t) &= \left(\frac{b}{\Omega}\right)e^{i(\omega-\alpha)t/2}\sin(\Omega t/2), \\ C_2(t) &= ie^{-i(\omega-\alpha)t/2}\left\{\cos(\Omega t/2) + i\left(\frac{\omega-\alpha}{\Omega}\right)\sin(\Omega t/2)\right\}, \end{aligned} \qquad (1.35)$$

with

$$\Omega = \left((\omega - \alpha)^2 + b^2\right)^{1/2}. \tag{1.36}$$

Thus

$$|C_1(t)|^2 = \left(\frac{b}{\Omega}\right)^2 \sin^2(\Omega t/2),$$

$$|C_2(t)|^2 = \cos^2(\Omega t/2) + \left(\frac{\omega - \alpha}{\Omega}\right)^2 \sin^2(\Omega t/2). \tag{1.37}$$

It can be seen that the normalization condition of equation (1.32) is satisfied by this set of solutions.

1.3 Density-matrix formulation of quantum mechanics

Measurable quantities in quantum mechanics are associated with probability, rather than with the probability amplitude. In section 1.1 and 1.2 the calculation of $|C_1(t)|^2$ and $|C_2(t)|^2$ from $C_1(t)$ and $C_2(t)$ was of eventual interest. We thus start by considering the two-by-two matrix defined as

$$\rho_{ij}(t) = C_i(t)C_j^*(t), \tag{1.38}$$

or

$$\rho(t) = \begin{pmatrix} C_1(t)C_1^*(t) & C_1(t)C_2^*(t) \\ C_2(t)C_1^*(t) & C_2(t)C_2^*(t) \end{pmatrix}. \tag{1.39}$$

This is known as the density matrix for the system. The density matrix in this particular case is formulated in the interaction picture. We can thus state the conservation of probability given in equation (1.32) as

$$\mathrm{Tr}(\rho) = 1, \quad \rho^2 = \rho. \tag{1.40}$$

Because it is Hermitian, we can diagonalize the density matrix of equation (1.38). The density matrix should then, when diagonalised, take the form

$$\rho_+ = \begin{pmatrix} 1 & 0 \\ 0 & 0 \end{pmatrix} \quad \text{or} \quad \rho_- = \begin{pmatrix} 0 & 0 \\ 0 & 1 \end{pmatrix}, \tag{1.41}$$

corresponding to the spin-up and spin-down states, respectively.

Next, let us demonstrate how it is possible to construct the equation of motion in terms of the density matrix without resorting to wave functions. Since the time derivative of ρ is

$$\dot{\rho}_{ij}(t) = \dot{C}_i(t)C_j^*(t) + C_i(t)\dot{C}_j^*(t), \tag{1.42}$$

for the system discussed in section 1.2, the differential equations of equation (1.31) for $C_i(t)$ lead to

$$\dot{\rho}_{11}(t) = -\dot{\rho}_{22}(t) = ib\left[\cos(\alpha t)\right]\left(e^{-i\omega t}\rho_{12} - e^{+i\omega t}\rho_{21}\right),$$
$$\dot{\rho}_{12}(t) = \dot{\rho}_{21}^{*}(t) = ib\left[\cos(\alpha t)\right]e^{i\omega t}(\rho_{11} - \rho_{22}). \tag{1.43}$$

The equation of motion can now be written in terms of the density matrix. If, for $\rho(t)$, the two-by-two matrix defined in equation (1.38) is used, the above set of equations can be written as

$$i\frac{\partial}{\partial t}\rho(t) = [G_I(t), \rho(t)], \tag{1.44}$$

where $G_I(t)$ is given in equation (1.28). In the interaction picture, this represents the Liouville equation for the density matrix.

It can be seen that finding the solution to this differential equation would be as difficult as the case in the interaction picture discussed in section 1.2. If, however, the rotating wave approximation is made, it should be possible to obtain a solution. Using this approximation we obtain

$$\dot{\rho}_{11}(t) = -\dot{\rho}_{22}(t) = i\frac{b}{2}\left(e^{-i(\omega-\alpha)t}\rho_{12} - e^{i(\omega-\alpha)t}\rho_{21}\right),$$
$$\dot{\rho}_{12}(t) = \dot{\rho}_{21}^{*}(t) = i\frac{b}{2}e^{i(\omega-\alpha)t}(\rho_{11} - \rho_{22}), \tag{1.45}$$

which are known as the optical Bloch equations. In order to find solutions to these equations, let us consider the following parametrization:

$$\rho_{11}(t) = a_{11}e^{\lambda t}, \quad \rho_{22} = a_{22}e^{\lambda t},$$
$$\rho_{12}(t) = a_{12}e^{i(\omega-\alpha)t}e^{\lambda t}, \tag{1.46}$$
$$\rho_{21}(t) = a_{21}e^{-i(\omega-\alpha)t}e^{\lambda t},$$

which leads to

$$\lambda a_{11} = -\lambda a_{22} = i\frac{b}{2}(a_{12} - a_{21}),$$

$$(\lambda + i(\omega - \alpha))a_{12} = i\frac{b}{2}(a_{11} - a_{22}), \tag{1.47}$$

$$(\lambda - i(\omega - \alpha))a_{21} = -i\frac{b}{2}(a_{11} - a_{22}).$$

Writing these equations in matrix form gives

$$\begin{pmatrix} \lambda & 0 & -ib/2 & ib/2 \\ 0 & \lambda & ib/2 & -ib/2 \\ -ib/2 & ib/2 & \lambda + i(\omega - \alpha) & 0 \\ ib/2 & -ib/2 & 0 & \lambda - i(\omega - \alpha) \end{pmatrix}\begin{pmatrix} a_{11} \\ a_{22} \\ a_{12} \\ a_{21} \end{pmatrix} = 0. \tag{1.48}$$

The quartic equation for λ is calculated to be

$$\lambda^2(\lambda^2 + (\omega - \alpha)^2 + b^2) = 0, \tag{1.49}$$

whose roots are

$$\lambda_1 = 0, \quad \lambda_2 = i\Omega, \quad \lambda_3 = -i\Omega. \tag{1.50}$$

Now, each element of $\rho(t)$ may be written as

$$\rho_{ij}(t) = a_{ij}^{(1)} + a_{ij}^{(2)}e^{i\Omega t} + a_{ij}^{(3)}e^{-i\Omega t}. \tag{1.51}$$

If we start with the initial condition $\rho_{11}(0) = \rho_{12}(0) = \rho_{21}(0)$ and $\rho_{22}(0) = 1$, we see that $\rho_{11}(t)$ may be written as

$$\rho_{11}(t) = \frac{K}{2}(1 - \cos \Omega t), \tag{1.52}$$

where the constant K is smaller than 1. In order to complete the solution, it is necessary to determine this constant using the optical Bloch equations of equation (1.45). We thus obtain

$$\rho_{22}(t) = 1 - \frac{K}{2} + \left(\frac{K}{2}\right)\cos \Omega t, \tag{1.53}$$

because $\rho_{11}(t) + \rho_{22}(t) = 1$. Additionally,

$$\dot{\rho}_{12}(t) = i\frac{b}{2}e^{i(\omega-\alpha)t}(K - 1 - K(\cos \Omega t)), \tag{1.54}$$

which is integrable. After the integration the result is

$$\rho_{12}(t) = \left(\frac{b}{2}\right)e^{i(\omega-\alpha)t}\left\{\left(\frac{K - 1}{\omega - \alpha}\right) + \frac{K}{b^2}[(\omega - \alpha)\cos \Omega t - i\Omega(\sin \Omega t)]\right\}. \tag{1.55}$$

We can from this determine that K is $(b/\omega)^2$ by using the initial condition $\rho_{12}(0) = 0$. To confirm the validity of this solution, it and its complex conjugate for ρ_{21} can be substituted into the first equation of equation (1.45) for $\dot{\rho}_{11}$. We see that the result is the same as that derived in section 1.2.

The density-matrix formalism of quantum mechanics has been used so far only to reproduce the result which is available in other pictures. No compelling reason therefore exists to choose this particular picture while abandoning others. In the case of calculating ensemble averages [12], however, the density matrix becomes an indispensable tool. For example, if we consider a system consisting of a statistical mixture of states with two different initial conditions, then we need to use the density matrix.

1.3.1 Mixed states

Because not all quantum states are in pure states, we discuss in this section a concrete example of a non-pure or mixed state. Using the eigenspinors applicable to a spin-1/2 particle we obtain for the θ ϕ-direction

$$u_+(\theta, \phi) = \begin{pmatrix} e^{-i\phi/2} \cos(\theta/2) \\ e^{i\phi/2} \sin(\theta/2) \end{pmatrix} \quad \text{and}$$

$$u_-(\theta, \phi) = \begin{pmatrix} -e^{-i\phi/2} \sin(\theta/2) \\ e^{i\phi/2} \cos(\theta/2) \end{pmatrix}$$

(1.56)

for the positive and negative directions, respectively. We can then define the polarization vector **P** as

$$P_i = (u_+(\theta, \phi))^\dagger \sigma_i u_+(\theta, \phi).$$

(1.57)

This will hereafter be written as $\langle \theta\phi | \sigma_i | \theta\phi \rangle$. We see then that

$$P_x = (\sin \theta)\cos \phi, \quad P_y = (\sin \theta)\sin \phi, \quad \text{and} \quad P_z = \cos \theta,$$

(1.58)

from which it is clear that this polarization vector has a unit length:

$$P_x^2 + P_y^2 + P_z^2 = 1.$$

(1.59)

If we consider the case of two independently prepared groups of particles, N_1 particles which have spin along the $\theta_1 \ \phi_1$-direction and N_2 particles with spin along the $\theta_2 \ \phi_2$-direction, then with

$$w_1 = N_1/(N_1 + N_2) \quad \text{and} \quad w_2 = N_2/(N_1 + N_2),$$

(1.60)

the polarization vector is the statistical average:

$$P_i = w_1\langle \theta_1\phi_1 | \sigma_i | \theta_1\phi_1 \rangle + w_2\langle \theta_2\phi_2 | \sigma_i | \theta_2\phi_2 \rangle.$$

(1.61)

This mixed-state polarization vector has the (magnitude)^2 of

$$P^2 = w_1^2 + w_2^2 + 2(w_1w_2)\cos \delta_{12},$$

(1.62)

where δ_{12} is the angle between the two directions. We see that P^2 is smaller than 1 and is greater than $(w_1 - w_2)^2$. If the two directions are the same, $P^2 = 1$, and if opposite to each other, then $P^2 = (w_1 - w_2)^2$.

For the eigenspinors of equation (1.56), the projection operator which selects only the spin states along the positive $\theta \ \phi$-direction is

$$\Lambda_+(\theta, \phi) = u_+(\theta, \phi)[u_+(\theta, \phi)]^\dagger.$$

(1.63)

This can be written in matrix form as

$$\Lambda_+(\theta, \phi) = \frac{1}{2}\begin{pmatrix} 1 + \cos \theta & e^{-i\phi} \sin \theta \\ e^{i\phi} \sin \theta & 1 - \cos \theta \end{pmatrix}.$$

(1.64)

If only the eigenspinors for the negative direction are selected, the projection operator is

$$\Lambda_-(\theta, \phi) = \frac{1}{2}\begin{pmatrix} 1 - \cos\theta & -e^{-i\phi}\sin\theta \\ -e^{i\phi}\sin\theta & 1 + \cos\theta \end{pmatrix}. \tag{1.65}$$

If the Pauli spin operator $\sigma(\theta, \phi) = (\sin\theta\cos\phi)\sigma_x + (\sin\theta\sin\phi)\sigma_y + (\cos\theta)\sigma_z$ is along the $\theta\,\phi$-direction, then it takes the form

$$\sigma(\theta, \phi) = \begin{pmatrix} \cos\theta & e^{-i\phi}\sin\theta \\ e^{i\phi}\sin\theta & -\cos\theta \end{pmatrix}. \tag{1.66}$$

The projection operators, in terms of this matrix, are

$$\Lambda_+(\theta, \phi) = \frac{1}{2}(I + \sigma(\theta, \phi)), \quad \text{and} \quad \Lambda_-(\theta, \phi) = \frac{1}{2}(I - \sigma(\theta, \phi)). \tag{1.67}$$

These operators are clearly Hermitian and satisfy

$$\Lambda_+(\theta, \phi) + \Lambda_-(\theta, \phi) = I. \tag{1.68}$$

From simple matrix multiplications we see that $(\Lambda_+(\theta, \phi))^2 = \Lambda_+(\theta, \phi)$ and $(\Lambda_-(\theta, \phi))^2 = \Lambda_-(\theta, \phi)$.

Let us define the density matrix for the mixed state as

$$\rho = w_1\Lambda_+(\theta, \phi) + w_2\Lambda_-(\theta, \phi), \tag{1.69}$$

which results in

$$\rho = \frac{I}{2} + \frac{1}{2}(w_1 - w_2)\sigma(\theta, \phi). \tag{1.70}$$

Because $\sigma(\theta, \phi)$ is traceless, we obtain $\mathrm{Tr}(\rho) = 1$, but ρ^2 is

$$\rho^2 = \frac{I}{4}\{1 + (w_1 - w_2)^2\} + \frac{1}{2}(w_1 - w_2)\sigma(\theta, \phi), \tag{1.71}$$

which is not equal to ρ. Indeed, ρ^2 is equal to ρ only for a pure state when $w_1 = 1$ and $w_2 = 0$, or vice versa. For mixed states, the trace of ρ^2 is less than 1.

By rotation we can diagonalize the density matrix of equation (1.70) because it is Hermitian. Since the diagonalized form of $\sigma(\theta, \phi)$ is σ_z, the diagonal form of the density matrix is

$$\rho_D = \frac{1}{2}\begin{pmatrix} 1 + w_1 - w_2 & 0 \\ 0 & 1 - w_1 + w_2 \end{pmatrix}. \tag{1.72}$$

If the system is fully polarized because $w_1 = 1$ and $w_2 = 0$, then the density matrix is

$$\rho_D = \begin{pmatrix} 1 & 0 \\ 0 & 0 \end{pmatrix}. \tag{1.73}$$

As the trace is invariant under rotations, trace properties of the density matrix are clearly demonstrated by these diagonal expressions.

We can also obtain the polarization vector through the relation

$$P_i = \mathrm{Tr}(\sigma_i \rho), \tag{1.74}$$

which after calculation gives

$$
\begin{aligned}
P_x &= (w_1 - w_2)(\sin\theta)\cos\phi, \\
P_y &= (w_1 - w_2)(\sin\theta)\sin\phi, \\
P_z &= (w_1 - w_2)\cos\theta.
\end{aligned}
\tag{1.75}
$$

The system is fully polarized when $w_1 = 1$ and $w_2 = 0$. However, if $w_1 = w_2 = 1/2$, the polarization vector vanishes. The magnitude of the polarization vector for this mixed state is

$$P^2 = P_x^2 + P_y^2 + P_z^2 = (w_1 - w_2)^2, \tag{1.76}$$

but becomes that of equation (1.59) for a pure state when $(w_1 - w_2)^2 = 1$.

The concrete example of the density matrix which we have worked out here is for a mixed state of spin-1/2 particles. It should be possible now to formulate a general theory of the density matrix.

1.3.2 Density matrix and ensemble average

Thus far only two-by-two density matrices have been discussed. However, the size of the matrix can be arbitrarily large, even being infinite-by-infinite. The index of the matrix can be continuous as, for example, the wave function $\psi(x)$ can be regarded as a column vector with a continuous index x.

For a particular quantum state which is a linear superposition of many different eigenstates ψ_n, we have

$$\psi(x) = \sum_n C_n \psi_n(x), \tag{1.77}$$

with

$$\sum_n |C_n|^2 = 1. \tag{1.78}$$

To obtain C_m for a particular m it is possible to use the projection operator

$$\Lambda_m = \psi_m(\psi_m)^\dagger. \tag{1.79}$$

This results in

$$\Lambda_m \psi(x) = C_m \psi_m(x), \tag{1.80}$$

with an understanding that

$$\Lambda_n \psi(x) = \psi_n(x)\big(\psi_n(x'), \psi(x')\big). \tag{1.81}$$

The projection operator is therefore seen to be a function of two variables x and x'. Thus it can be written as $\Lambda_n(x, x')$ whenever necessary with x and x' as the continuous matrix indices. This Hermitian operator has the property that

$$\Lambda_n \Lambda_m = \delta_{nm} \Lambda_m, \quad \sum_n \Lambda_n = I, \tag{1.82}$$

with I being the identity operator.

We can then define the density matrix as

$$\rho(x, x') = \sum_n |C_n|^2 \Lambda_n(x, x'), \tag{1.83}$$

so that for a given operator $A(x)$, $\text{Tr}(\rho A)$ is

$$\text{Tr}(\rho A) = \sum_n |C_n|^2 (\psi_n, A\psi_n). \tag{1.84}$$

The expectation value is

$$\langle A \rangle = \sum_n \sum_m C_m^* C_n (\psi_m, A\psi_n), \tag{1.85}$$

demonstrating that equation (1.84) is quite different from equation (1.85). We can write the difference between the two quantities as

$$\langle A \rangle - \text{Tr}(\rho A) = \sum_{n \neq m} C_m^* C_n (\psi_m, A\psi_n). \tag{1.86}$$

Can this quantity vanish? We know that coefficient $C_m^* C_n$ is real and positive when $m = n$, but complex otherwise, and carries a phase factor. This means that when $n \neq m$ and the ensemble average with random phase factors is calculated, the average of each $C_m^* C_n$ vanishes.

As an example, we shall discuss the one-dimensional harmonic oscillator, for which the Schrödinger equation is

$$i\frac{\partial}{\partial t}\psi(x, t) = \frac{1}{2}\left\{ m\omega^2 x^2 - \frac{1}{m}\left(\frac{\partial}{\partial x}\right)^2 \right\}\psi(x, t). \tag{1.87}$$

With

$$\big(\psi(x, t), \psi(x, t)\big) = \sum_n |C_n|^2 = 1, \tag{1.88}$$

the most general form of the normalized solution is

$$\psi(x, t) = e^{-i\omega t/2} \sum_n C_n e^{-in(\omega t)}\psi_n(x). \tag{1.89}$$

In the above $\psi_n(x)$ is the solution of the time-independent equation:

$$\frac{1}{2}\left(m\omega^2 x^2 - \frac{1}{m}\left(\frac{\partial}{\partial x}\right)^2\right)\psi_n(x) = \omega(n + 1/2)\psi_n(x). \tag{1.90}$$

Then we can write the expectation value $\langle A \rangle = (\psi(x, t), A\psi(x, t))$ of an operator $A(x)$ as

$$\langle A \rangle = \sum_n |C_n|^2 \left(\psi_n(x), A(x)\psi_n(x)\right)$$
$$+ \sum_{n \neq m} C_m^* C_n e^{i\omega(m-n)t} \left(\psi_m(x), A(x)\psi_n(x)\right). \tag{1.91}$$

The second term above vanishes when we take the time average of this quantity. Taking the ensemble average for many particles with different phases of C_n, produces the same result. As a consequence, this gives the ensemble average as

$$\overline{\langle A \rangle} = \sum_n |C_n|^2 \left(\psi_n(x), A(x)\psi_n(x)\right). \tag{1.92}$$

This problem can be treated very conveniently if the density matrix defined as

$$\rho(x, x') = \sum_n |C_n|^2 \psi_n(x)\psi_n^*(x'), \tag{1.93}$$

is introduced. Then we can write

$$\overline{\langle A \rangle} = \int dx' \int A(x', x)\rho(x, x')dx, \tag{1.94}$$

with

$$A(x', x) = \delta(x' - x)A(x). \tag{1.95}$$

Often written as

$$\overline{\langle A \rangle} = \text{Tr}(\rho A), \tag{1.96}$$

the above expression is seen to be the trace of the matrix $\rho(x', x)A(x, x')$.

When $C_n = \delta_{nm}$ for a given value of m, it is clear that the system is in a pure state; otherwise it is in a mixed state. We know that $\sum_n |C_n|^2 = 1$ and $\text{Tr}(\rho) = 1$. Then we can determine that $\text{Tr}(\rho^2) = \sum_n |C_n|^4$ which is 1 for a pure state and is less than 1 for mixed states.

The harmonic oscillator in thermal equilibrium [4] has the density matrix

$$\rho_T(x, x') = (1 - e^{-\omega/k_B T}) \sum_n e^{-n\omega/k_B T} \psi_n(x)\psi_n^*(x'), \tag{1.97}$$

where k_B is Boltsmann's constant. For demonstrating the above, the following is probably the best known example. Starting from equation (1.97), $\text{Tr}(\rho)$ is

$$\text{Tr}(\rho_T) = \int \rho_T(x, x)dx = (1 - e^{-\omega/k_B T}) \sum_n e^{-n\omega/k_B T}, \tag{1.98}$$

which, after summation, becomes 1. From this $\mathrm{Tr}((\rho_T)^2)$ is then

$$\mathrm{Tr}(\rho_T\rho_T) = \int \left\{ \int \rho_T(x, x')\, \rho_T(x', x)dx' \right\} dx. \tag{1.99}$$

Using equation (1.97) we obtain

$$\mathrm{Tr}\big((\rho_T)^2\big) = (1 - e^{-\omega/k_BT})^2 \sum_n e^{-2n\omega/k_BT} = \tanh(\omega/2k_BT), \tag{1.100}$$

which is less than 1, and becomes 1 only when T becomes 0. The system is in a pure state when $T = 0$. This is the ground-state of the harmonic oscillator.

1.3.3 Time dependence of the density matrix

In section 1.3.2 we discussed the density matrix for the harmonic oscillator and, although we started with a time-dependent system, we ended up with a time-independent density matrix. Does this tell us that the density matrix is intrinsically time-independent? Clearly the answer is no, as the example discussed in the initial part of section 1.3 leads to a time-dependent density matrix.

For time-dependent wave functions, the density matrix is written as

$$\rho(x, x', t) = \sum_n w_n \psi_n(x, t)\psi_n^*(x', t). \tag{1.101}$$

Then differentiation of this equation with respect to time results in

$$\begin{aligned}
i\frac{\partial}{\partial t}\rho(x, x', t) &= \sum_n w_n \left\{ i\frac{\partial}{\partial t}\psi_n(x, t)\psi_n^*(x', t) \right. \\
&\quad \left. - \psi_n(x, t)i\frac{\partial}{\partial t}\psi_n^*(x', t) \right\} \\
&= \sum_n w_n \big\{ H\psi_n(x, t)\psi_n^*(x', t) \\
&\quad - \big(\psi_n(x, t)\big)H\psi_n^*(x', t) \big\}.
\end{aligned} \tag{1.102}$$

Thus, in the Schrödinger picture, the density matrix satisfies the Liouville equation:

$$i\frac{\partial}{\partial t}\rho(x, x', t) = [H, \rho(x, x', t)]. \tag{1.103}$$

Since the Hamiltonian is independent of time in the Schrödinger picture, the solution formally takes the form

$$\rho(t) = e^{-iHt}\rho(0)e^{iHt}. \tag{1.104}$$

Only when the Hamiltonian commutes with $\rho(0)$ is the density matrix independent of time.

For the one-dimensional harmonic oscillator the density matrix has the form

$$\rho(t) = e^{-iHt}\rho(0)e^{iHt} = \sum_n w_n e^{-iHt}\psi_n(x, t)e^{iHt}\psi_n^*(x', t). \tag{1.105}$$

Since the wave function $\psi_n(x, t)$ is an eigenstate of H, the density matrix is independent of time. The density matrix commutes with the Hamiltonian.

As an example where the Hamiltonian and the density matrix do not commute, we consider the case of a spin-1/2 particle in a constant magnetic field. Equation (1.3) gives the Hamiltonian for this system, and equation (1.4) gives the wave function. This wave function is not an energy eigenstate. The density matrix is given by

$$\rho(t) = \begin{pmatrix} aa^* & ab^*e^{-\omega t} \\ a^*be^{i\omega t} & bb^* \end{pmatrix}. \tag{1.106}$$

The time evolution, in the Schrödinger picture, indeed, can be written in the form of equation (1.20) as

$$\rho(t) = \begin{pmatrix} e^{-i\omega t/2} & 0 \\ 0 & e^{i\omega t/2} \end{pmatrix}\begin{pmatrix} aa^* & ab^* \\ a^*b & bb^* \end{pmatrix}\begin{pmatrix} e^{i\omega t/2} & 0 \\ 0 & e^{-i\omega t/2} \end{pmatrix}. \tag{1.107}$$

Unless $a = 1$ and $b = 0$, or $a = 0$ and $b = 1$, this matrix does not commute with the Hamiltonian. The density matrices are

$$\begin{pmatrix} 1 & 0 \\ 0 & 0 \end{pmatrix} \text{ and } \begin{pmatrix} 0 & 0 \\ 0 & 1 \end{pmatrix}, \tag{1.108}$$

respectively, for those special cases. The spin then is along the positive and negative directions, respectively. Since the spin direction does not change in time, the density matrix is also independent of time.

Suppose instead that the spin is along the x-direction at $t = 0$, then the density matrix is given by

$$\rho_x(t) = \frac{1}{2}\begin{pmatrix} 1 & e^{-i\omega t} \\ e^{i\omega t} & 1 \end{pmatrix}. \tag{1.109}$$

The expectation values of σ_i can be calculated by using the formula $\langle\sigma_i\rangle = \mathrm{Tr}(\rho_x\sigma_i)$. The trace calculation will lead to the same result as that expected from equation (1.8), namely, $\langle\sigma_z\rangle = 0$, $\langle\sigma_x\rangle = \cos\omega t$, and $\langle\sigma_y\rangle = \sin\omega t$.

Suppose now that the spin is along the negative x-direction at $t = 0$, then the density matrix is given by

$$\rho_{-x}(t) = \frac{1}{2}\begin{pmatrix} 1 & -e^{-i\omega t} \\ -e^{i\omega t} & 1 \end{pmatrix}. \tag{1.110}$$

It can be seen from equations (1.109) and (1.110) that $\rho_x(t)$ and $\rho_{-x}(t)$ are pure-state density matrices, which satisfy the conditions $\mathrm{Tr}(\rho) = 1$ and $\rho^2 = \rho$.

If we have a system of N spin-1/2 particles, where $N/2$ particles have spin along the x-direction, and $N/2$ particles have spin along the negative x-direction, then the ensemble average of the density matrix is

$$\rho = \begin{pmatrix} 1/2 & 0 \\ 0 & 1/2 \end{pmatrix}. \tag{1.111}$$

This is not a pure-state density matrix, because although it satisfies the trace relation $\text{Tr}(\rho) = 1$, the $\text{Tr}(\rho^2)$ is less than 1.

We saw in the interaction picture that it is possible to derive the Liouville equation as the example in the beginning of section 1.3 illustrates. We continue now the discussion presented there on the rotating wave approximation. For the spin initially along the z-direction,

$$\rho_{11}(t) = 1 - \rho_{22}(t) = \left(\frac{b}{\Omega}\right)^2 (\sin(\Omega t/2))^2,$$

$$\rho_{12}(t) = \rho_{21}^*(t) = -i\left(\frac{b}{\Omega}\right) e^{i(\omega-\alpha)t}(\sin(\Omega t/2)) \tag{1.112}$$

$$\times \left\{ \cos(\Omega t/2) - i\left(\frac{\omega - \alpha}{\Omega}\right) \sin(\Omega t/2) \right\}.$$

However, if the spin is along the negative z-axis,

$$\rho_{11}(t) = 1 - \rho_{22}(t) = 1 - \left(\frac{b}{\Omega}\right)^2 (\sin(\Omega t/2))^2,$$

$$\rho_{12}(t) = \rho_{21}^*(t) = i\left(\frac{b}{\Omega}\right) e^{i(\omega-\alpha)t}(\sin(\Omega t/2)) \tag{1.113}$$

$$\times \left\{ \cos(\Omega t/2) - i\left(\frac{\omega - \alpha}{\Omega}\right) \sin(\Omega t/2) \right\}.$$

We see that the diagonal form given in equation (1.111) is again reproduced by the ensemble average of these two initial conditions.

By noting that the formula for the measurable quantity

$$\langle A \rangle = \text{Tr}(\rho A) \tag{1.114}$$

is invariant under unitary transformation of the operators and the density matrix, it is possible to study the expectation values of operators. This expression, valid in the interaction picture, is also valid in the Schrödinger and Heisenberg pictures. The density matrices of equations (1.112) and (1.113) have been computed in the interaction picture.

In order to calculate the expectation value of $\langle \sigma_i \rangle$, it is necessary to use the operators in the interaction picture, where

$$\sigma_x(t) = \begin{pmatrix} e^{i\omega t/2} & 0 \\ 0 & e^{-i\omega t/2} \end{pmatrix} \begin{pmatrix} 0 & 1 \\ 1 & 0 \end{pmatrix} \begin{pmatrix} e^{-i\omega t/2} & 0 \\ 0 & e^{i\omega t/2} \end{pmatrix},$$

$$= \begin{pmatrix} 0 & e^{i\omega t} \\ e^{-i\omega t} & 0 \end{pmatrix}. \tag{1.115}$$

Similarly,

$$\sigma_y(t) = \begin{pmatrix} 0 & -ie^{i\omega t} \\ ie^{-i\omega t} & 0 \end{pmatrix} \quad \text{and} \quad \sigma_z(t) = \begin{pmatrix} 1 & 0 \\ 0 & -1 \end{pmatrix}. \tag{1.116}$$

Here $\sigma_x(t)$, $\sigma_y(t)$, and $\sigma_z(t)$ are the Pauli spin matrices in the interaction picture. We can now calculate the polarization vector from

$$P_x(t) = \mathrm{Tr}\,(\rho(t)\sigma_x(t)) = 2\{\mathrm{Re}\,(e^{-i\omega t}\rho_{12}(t))\},$$
$$P_y(t) = \mathrm{Tr}\,(\rho(t)\sigma_y(t)) = 2\{\mathrm{Im}\,(e^{-i\omega t}\rho_{12}(t))\}, \tag{1.117}$$
$$P_z(t) = \mathrm{Tr}\,(\rho(t)\sigma_z(t)) = \rho_{11}(t) - \rho_{22}(t).$$

These calculations were carried out in the interaction picture. However as

$$P_i(t) = \mathrm{Tr}\,(\rho(t)\sigma_i(t)) = \mathrm{Tr}\,(e^{-iHt}\rho(0)e^{iHt}e^{-iHt}\sigma_i(0)e^{iHt}), \tag{1.118}$$

if the calculation were done in the Schrödinger or Heisenberg picture the result would be the same. Since it is clear that $P_i(t)$ is a measurable quantity, and the calculation is independent of the picture of quantum mechanics chosen, we confirm there is only one quantum mechanics.

When the spin is initially along the z-direction, we can use the $\rho(t)$ matrix given in equation (1.112) to calculate the polarization vector:

$$P_x(t) = \frac{2b\sin(\Omega t/2)}{\Omega}\left\{(\cos(\Omega t/2))\sin(\alpha t) + \left(\frac{\omega - \alpha}{\Omega}\right)(\sin(\Omega t/2))\cos(\alpha t)\right\},$$

$$P_y(t) = \frac{2b\sin(\Omega t/2)}{\Omega}\left\{(\cos(\Omega t/2))\cos(\alpha t) - \left(\frac{\omega - \alpha}{\Omega}\right)(\sin(\Omega t/2))\sin(\alpha t)\right\}, \tag{1.119}$$

$$P_z(t) = 1 - 2\left(\frac{b}{\Omega}\right)^2(\sin(\Omega t/2))^2.$$

We see that the magnitude of this polarization vector can be measured from

$$(P_x(t))^2 + (P_y(t))^2 + (P_z(t))^2. \tag{1.120}$$

If we use the polarization vector given in equation (1.119), this quantity is one and independent of time.

When instead, the solution for $\rho(t)$ given in equation (1.113) is used, then the polarization vector is the negative of the vector given in equation (1.119). This is because the spin is initially along the negative z-direction. However, it is still a pure state. By mixing states with two different initial polarizations, as was done in section 1.3.1, we find

$$(P_x(t))^2 + \left(P_y(t)\right)^2 + (P_z(t))^2 = (w_1 - w_2)^2 \tag{1.121}$$

which is also independent of time.

1.4 Further contents of Heisenberg's commutation relations

Let us write Heisenberg's commutation relations again,

$$[x_i, p_j] = i\delta_{ij}. \tag{1.122}$$

In physics, these relations are called the uncertainty relations, and the present forms of quantum mechanics are built upon these relations. This does not prevent us from constructing further mathematical instruments from these commutation relations. Among others is the Lie algebra of the three-dimensional rotation group. The algebra of the creation and annihilation operators is also a direct consequence of this algebra. Let us examine them in detail.

1.4.1 Rotation group and its extension to the Lorentz group

The dynamical equations in quantum mechanics are all consistent with Heisenberg's uncertainty relation, where we start with the three coordinate variables x_i and three operators $p_j = -i\frac{\partial}{\partial x_j}$.

From these, we can construct the operators

$$L_i = \epsilon_{ijk} x_j p_k. \tag{1.123}$$

These operators satisfy the commutation relations

$$\left[L_i, L_j\right] = i\epsilon_{ijk} L_k. \tag{1.124}$$

Most certainly, this set of commutation relations is another form of the Heisenberg relation of equation (1.122). On the other hand, this set is also widely known as the Lie algebra of the three-dimensional rotation group.

It is also well known that the two-by-two matrices

$$J_i = \frac{1}{2}\sigma_i, \tag{1.125}$$

where σ_i are the Pauli spin matrices, satisfy the same set of commutation relations as that for the rotation group:

$$\left[J_i, J_j\right] = i\epsilon_{ijk} J_k. \tag{1.126}$$

The two-by-two Pauli matrices are Hermitian and traceless. We can next consider three additional matrices defined as

$$K_i = \frac{i}{2}\sigma_i. \tag{1.127}$$

These three matrices do not lead to a closed set of commutation relations. Instead, the commutator of two K_i matrices leads to

$$\left[K_i, K_j\right] = -i\epsilon_{ijk}J_k, \tag{1.128}$$

and

$$\left[J_i, K_j\right] = i\epsilon_{ijk}K_k. \tag{1.129}$$

Thus, the six traceless two-by-two matrices, namely J_i and K_i, satisfy the closed set of commutation relations consisting of equations (1.126), (1.128), and (1.129).

As we shall see in chapter 2, this set of commutation relations leads to the group of Lorentz transformations. There are four complex elements in one two-by-two matrix. If they are to be traceless, there are only six independent parameters. Thus, these six matrices constitute the complete set of generators for the most general form of two-by-two unimodular matrices. If the determinant of a given matrix is one, this matrix is said to be unimodular.

There are many two-by-two matrices in optical and information sciences. Some of them are generated by the Pauli spin matrices, and some of them are not. In either case, they are all generated by the J_i and K_i matrices. It is interesting to note that the commutation relations of equation (1.126) are another form of Heisenberg commutation relations. If we add the anti-Hermitian operators K_i, the resulting set of relations form the Lie algebra for the Lorentz group. By Lie algebra we mean the closed set of commutation relations for the generators of the group. The relations given in equation (1.126) constitute the Lie algebra for the rotation group.

1.4.2 Harmonic oscillators and Fock space

There is another useful algebra derivable from the Heisenberg commutation relations. We can consider the operators

$$a_i = \frac{1}{\sqrt{2}}(x_i + ip_i) = \frac{1}{\sqrt{2}}\left(x_i + \frac{\partial}{\partial x_i}\right),$$

$$a_i^\dagger = \frac{1}{\sqrt{2}}(x_i - ip_i) = \frac{1}{\sqrt{2}}\left(x_i - \frac{\partial}{\partial x_i}\right). \tag{1.130}$$

Then

$$\left[a_i, a_j^\dagger\right] = \delta_{ij} \quad \text{and} \quad [a_i, a_j] = 0. \tag{1.131}$$

This set of commutators is another way of expressing the Heisenberg relations given in equation (1.1).

Let us consider the one-dimensional harmonic oscillator, and drop the subscripts i and j. We thus write equation (1.130) as

$$a = \frac{1}{\sqrt{2}}(x + ip) = \frac{1}{\sqrt{2}}\left(x + \frac{\partial}{\partial x}\right),$$
$$a^\dagger = \frac{1}{\sqrt{2}}(x - ip) = \frac{1}{\sqrt{2}}\left(x - \frac{\partial}{\partial x}\right).$$

(1.132)

We can next introduce the harmonic oscillator differential equation

$$\frac{1}{2}\left\{-\left(\frac{\partial}{\partial x}\right)^2 + x^2\right\}\chi_n(x) = \left(n + \frac{1}{2}\right)\chi_n(x).$$

(1.133)

Its solution is the wave function in the nth excited state, which is written as

$$\chi_n(x) = \frac{1}{\sqrt{\pi 2^n n!}} H_n(x) e^{-x^2/2},$$

(1.134)

where $H_n(x)$ is the Hermite polynomial. If we are only interested in the quantum number n, it is more convenient to write this wave function as $|n\rangle$. Then

$$a|n\rangle = \sqrt{n}\,|n - 1\rangle, \quad \text{and} \quad a^\dagger|n\rangle = \sqrt{n + 1}\,|n + 1\rangle.$$

(1.135)

Thus, a and $a^\dagger$ are called the step-down and step-up operators, respectively, for the harmonic oscillator wave functions.

This algebra is the starting point for the Fock space [5] where the number n serves as the number of particles or photons in a given state. The operator a reduces the number of particles by one, and $a^\dagger$ adds one. They are the annihilation and creation operators, respectively. These operators are the basic mathematical device for quantum optics.

1.4.3 Dirac's two-oscillator system

In his paper published in 1963, Paul A M Dirac [3] considered two oscillators, with a_1, $a_1^\dagger$ and a_2, $a_2^\dagger$. He then noticed that the following combinations are possible:

$$a_i a_j, \quad a_i^\dagger a_j^\dagger, \quad a_i^\dagger a_j, \quad \text{and} \quad a_i a_j^\dagger.$$

(1.136)

There are thus sixteen combinations. The two-mode coherent state is generated by a set of these bilinear combinations [13]. In chapter 4 we explicitly derive these sixteen combinations in terms of Dirac spinors.

In 1963 Dirac showed [3] that it is possible to choose ten sets of these bilinear combinations which satisfy the closed set of commutation relations for the generators of the $O(3, 2)$ group or the Lorentz group applicable to three space-like and two time-like coordinates. This group, also known as the $3 + 2$ de Sitter group, can be regarded as two coupled Lorentz groups. Thus, it is possible to say that the symmetry of the Lorentz group is derivable from Heisenberg's commutation relations.

It was shown later [6] that all sixteen of those bilinear combinations lead to the Lie algebra of the $O(3, 3)$ group or the Lorentz group applicable to three space-like and three time-like coordinates. We shall discuss in chapter 7 what useful roles these $O(3, 2)$ and $O(3, 3)$ groups play in optical and information sciences.

References

[1] Başkal S, Kim Y S, and Noz M E 2019 Poincaré symmetry from Heisenberg's Uncertainty Relations *Symmetry* **11** 409

[2] Blum K 2012 *Density Matrix Theory and Applications Springer Series on Atomic, Optical and Plasma Physics* vol 64 3rd edn (Heidelberg: Springer)

[3] Dirac P A M 1963 A remarkable representation of the 3 + 2 de Sitter group *J. Math. Phys.* **4** 901–9

[4] Feynman R P, Kislinger M, and Ravndal F 1971 Current matrix elements from a relativistic quark model *Phys. Rev.* D **3** 2706–32

[5] Fock V 1934 Quanten elecktrodynamik *Phys. Z. Sowjetunion* **6** 425–69

[6] Han D, Kim Y S, and Noz M E 1995 *O* (3,3)–like symmetries of coupled harmonic oscillators *J. Math. Phys.* **36** 3940–54

[7] Itzykson C and Zuber J B 2005 *Quantum Field Theory* (*Dover Books on Physics*) (Mineola, NY: Dover)

[8] Kim Y S and Noz M E 1991 *Phase Space Picture of Quantum Mechanics: Group Theoretical Approach* (*Lecture Notes in Physics* vol 40) (Singapore: World Scientific)

[9] Loudon R 2000 *The Quantum Theory of Light* 3rd edn (Oxford: Oxford University Press)

[10] Pantell R H and Puthoff H E 1969 *Fundamentals of Quantum Electronics* (New York: Wiley)

[11] Shirley J H 1965 Solution of the Schrödinger equation with a Hamiltonian periodic in time *Phys. Rev.* B **138** B979–87

[12] Von Neumann J, Beyer R T, and Wheeler N A 2018 *Mathematical Foundations of Quantum Mechanics* (Princeton, NJ: Princeton University Press)

[13] Yuen H P 1976 Two-photon coherent states of the radiation field *Phys. Rev.* A **13** 2226–43

IOP Publishing

Mathematical Devices for Optical Sciences

Sibel Başkal, Young S Kim, and Marilyn E Noz

Chapter 2

Lorentz group and its representations

The rotation group and its representations are quite familiar to us in dealing with rotations in three-dimensional space, particularly in atomic physics and radiative atomic transitions [18, 20], as well as in quantum optics. If additionally, we combine the equally familiar Lorentz boost with the rotation group, the result is the Lorentz group.

Since both of these groups are continuous they are referred to as *Lie groups* [6]. For these groups, it is convenient to use their generators. For the three-dimensional rotation group, there are three generators, and they form a closed set of commutation relations. This closed set is called the *Lie algebra* [6]. When the rotation group is augmented with the Lorentz boost, the result is the Lorentz group. This group has six generators and its Lie algebra consists of a closed set of commutation relations among these generators. This group generates Lorentz transformations applicable to the four-dimensional Minkowskian space consisting of three space-like dimensions and one time-like dimension.

The transformation matrices of this group are four-by-four, with six generators. As in the case of the rotation group, the Lorentz group can be generated by six two-by-two matrices. Three of them are Hermitian and generate the rotation group known as $SU(2)$. The remaining three are anti-Hermitian and generate two-by-two matrices corresponding to Lorentz boosts in three different directions. The group generated by these Hermitian and anti-Hermitian generators is called $SL(2, c)$ in the literature.

There are many two-by-two matrices in the current literature on optics and information theory. Not all of them are rotation matrices. On the other hand, they can be dealt with within the frame of $SL(2, c)$ corresponding to the group of Lorentz transformations. In this way, the Lorentz group, used by Einstein for fast-moving particles, can serve as one of the basic languages in optics, quantum information theory, and entanglement problems, as is illustrated in figure 2.1.

Hence in section 2.1 we introduce the group of four-by-four matrices which perform Lorentz transformations on this Minkowskian space, including their

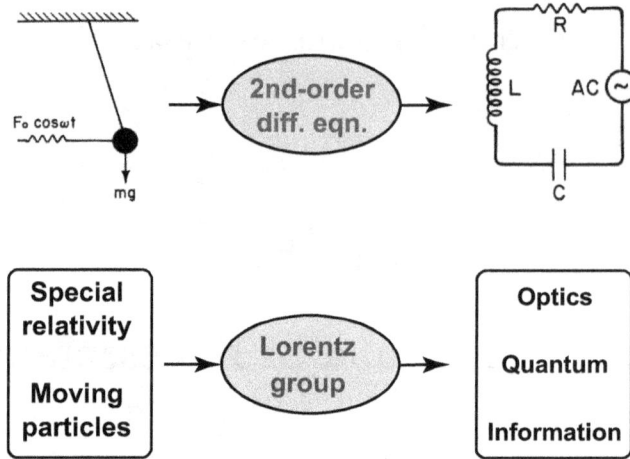

Figure 2.1. The damped harmonic oscillator leads to a second-order differential equation [8]. This equation serves as the mathematical instrument for understanding resonance circuits in electronics. Likewise, Einstein had to use the Lorentz group for fast-moving particles. We shall study in this book how this mathematical device serves as one of the basic languages for optics, and the foundation of quantum mechanics, information theory [15], and entanglement problems [3].

generators and their Lie algebra. In section 2.2 it is shown that the group $SL(2, c)$, the group of two-by-two unimodular matrices, shares the same algebra as the Lorentz group. It is thus easier to study the Lorentz group using these two-by-two matrices. In section 2.3 the four-component space–time and momentum–energy four-vectors are written in the form of two-by-two matrices.

In section 2.4 we study the transformation properties of the $SL(2, c)$ matrices and those that are applicable to the four-vector. They are different. In section 2.5 we study the subgroups of $SL(2, c)$, thus those of the Lorentz group. Of particular interest are the rotation subgroup, the $Sp(2)$ subgroup consisting of transformation matrices with real parameters, and $O(2, 1)$, which is the Lorentz subgroup in two space dimensions and one time dimension.

In section 2.6, it is shown that the $Sp(2)$ group with three independent parameters can be written as one boost matrix sandwiched between two rotation matrices. It is then shown that this becomes a triangular matrix which cannot be diagonalized. In section 2.7, it is noted the bilinear conformal transformation can be used for transformations within the $SL(2, c)$ group.

2.1 Lie algebra of the Lorentz group

If we start with the coordinates x, y, and z in three-dimensional space, we can rotate this coordinate system around the z-axis by writing

$$\begin{pmatrix} \cos\phi & -\sin\phi & 0 \\ \sin\phi & \cos\phi & 0 \\ 0 & 0 & 1 \end{pmatrix} \begin{pmatrix} x \\ y \\ z \end{pmatrix} = \begin{pmatrix} (\cos\phi)x - (\sin\phi)y \\ (\sin\phi)x + (\cos\phi)y \\ z \end{pmatrix}. \tag{2.1}$$

While z remains unchanged, x and y become $(\cos \phi)x - (\sin \phi)y$ and $(\sin \phi)x + (\cos \phi)y$, respectively. The matrix above can also be written as

$$e^{-i\phi L_3} = \sum_n \frac{(-i\phi L_3)^n}{n!}, \tag{2.2}$$

with

$$L_3 = \begin{pmatrix} 0 & -i & 0 \\ i & 0 & 0 \\ 0 & 0 & 0 \end{pmatrix}. \tag{2.3}$$

This matrix represents the generator of rotations around the z-axis. Likewise, generators of rotations around the x-axis and also around the y-axis can be written as

$$L_1 = \begin{pmatrix} 0 & 0 & 0 \\ 0 & 0 & -i \\ 0 & i & 0 \end{pmatrix} \quad \text{and} \quad L_2 = \begin{pmatrix} 0 & 0 & i \\ 0 & 0 & 0 \\ -i & 0 & 0 \end{pmatrix}, \tag{2.4}$$

respectively.

These generators satisfy the commutation relations

$$[L_1, L_2] = iL_3, \quad [L_2, L_3] = iL_1, \quad \text{and} \quad [L_3, L_1] = iL_2, \tag{2.5}$$

which can be succinctly expressed as

$$\left[L_i, L_j \right] = i\epsilon_{ijk}L_k. \tag{2.6}$$

This closed set of commutation relations is called the *Lie algebra* of the three-dimensional rotation group.

There are other forms of operators that satisfy the same Lie algebra. The three operators

$$L_i = -i\epsilon_{ijk}\left(x_j \frac{\partial}{x_k} - x_k \frac{\partial}{x_j} \right), \tag{2.7}$$

which satisfy the Lie algebra given in equation (2.5) or equation (2.6), are applicable to functions of the coordinate variables x, y, and z.

As for the Lorentz boost, Einstein formulated how events appear to observers moving with respect to each other. For example, if one observer is stationary and the other is moving with the velocity v along the z-direction (ignoring the transverse components), their coordinates are related by

$$z' = \gamma(z + \beta t). \tag{2.8}$$

The time variable also undergoes a change, so that

$$t' = \gamma(t + \beta z), \tag{2.9}$$

where $\gamma = (1 - \beta^2)^{-\frac{1}{2}}$, $\beta = v/c$, and c is the speed of light. Combining equations (2.8) and (2.9) in all four dimensions leads to the transformation law which can be written as

$$
\begin{pmatrix} t' \\ z' \\ x' \\ y' \end{pmatrix} = \begin{pmatrix} \cosh\eta & \sinh\eta & 0 & 0 \\ \sinh\eta & \cosh\eta & 0 & 0 \\ 0 & 0 & 1 & 0 \\ 0 & 0 & 0 & 1 \end{pmatrix} \begin{pmatrix} t \\ z \\ x \\ y \end{pmatrix},
\tag{2.10}
$$

where $\beta = \tanh\eta$. We shall hereafter use the convention $c = 1$. We shall also use the coordinate system $(t, z, x, y,)$ in Minkowskian space since the Lorentz boost will usually be made along the z-direction.

The generator for the boost along the z-direction is

$$
K_3 = \begin{pmatrix} 0 & i & 0 & 0 \\ i & 0 & 0 & 0 \\ 0 & 0 & 0 & 0 \\ 0 & 0 & 0 & 0 \end{pmatrix}.
\tag{2.11}
$$

Likewise, we can consider boosts along the x- and y-directions, and their generators are

$$
K_1 = \begin{pmatrix} 0 & 0 & i & 0 \\ 0 & 0 & 0 & 0 \\ i & 0 & 0 & 0 \\ 0 & 0 & 0 & 0 \end{pmatrix} \quad \text{and} \quad K_2 = \begin{pmatrix} 0 & 0 & 0 & i \\ 0 & 0 & 0 & 0 \\ 0 & 0 & 0 & 0 \\ i & 0 & 0 & 0 \end{pmatrix},
\tag{2.12}
$$

respectively. The generators applicable to boosts defined over the Minkowskian space are thus

$$
K_i = -i\left(x_i \frac{\partial}{\partial t} + t \frac{\partial}{\partial x_i} \right).
\tag{2.13}
$$

In this Minkowskian space, the rotation matrix of equation (2.1) is expanded to the four-by-four matrix

$$
\begin{pmatrix} 1 & 0 & 0 & 0 \\ 0 & 1 & 0 & 0 \\ 0 & 0 & \cos\phi & -\sin\phi \\ 0 & 0 & \sin\phi & \cos\phi \end{pmatrix} \begin{pmatrix} t \\ z \\ x \\ y \end{pmatrix} = \begin{pmatrix} t \\ z \\ (\cos\phi)x - (\sin\phi)y \\ (\sin\phi)x + (\cos\phi)y \end{pmatrix},
\tag{2.14}
$$

and its generator of equation (2.3) becomes the four-by-four matrix

$$
J_3 = \begin{pmatrix} 0 & 0 & 0 & 0 \\ 0 & 0 & 0 & 0 \\ 0 & 0 & 0 & -i \\ 0 & 0 & i & 0 \end{pmatrix}.
\tag{2.15}
$$

Likewise, the three-by-three matrices of L_1 and L_2 become expanded to

$$J_1 = \begin{pmatrix} 0 & 0 & 0 & 0 \\ 0 & 0 & 0 & i \\ 0 & 0 & 0 & 0 \\ 0 & -i & 0 & 0 \end{pmatrix} \quad \text{and} \quad J_2 = \begin{pmatrix} 0 & 0 & 0 & 0 \\ 0 & 0 & -i & 0 \\ 0 & i & 0 & 0 \\ 0 & 0 & 0 & 0 \end{pmatrix}, \tag{2.16}$$

respectively. These four-by-four matrices satisfy the same Lie algebra given for the three-by-three matrices in equation (2.6):

$$\left[J_i, J_j \right] = i\epsilon_{ijk} J_k. \tag{2.17}$$

If we return to the boost generators of equations (2.11) and (2.12), and take commutation relations then

$$\left[K_i, K_j \right] = -i\epsilon_{ijk} J_k. \tag{2.18}$$

It is clear that the three boost generators cannot form a closed set of commutation relations. However, if we take commutation relations between the J and K matrices we obtain

$$\left[J_i, K_j \right] = i\epsilon_{ijk} K_k. \tag{2.19}$$

The four-by-four J and K matrices are called the generators of the Lorentz group. They form the following closed set of commutation relations,

$$\left[J_i, J_j \right] = i\epsilon_{ijk} J_k, \quad \left[J_i, K_j \right] = i\epsilon_{ijk} K_k, \quad \text{and} \quad \left[K_i, K_j \right] = -i\epsilon_{ijk} J_k, \tag{2.20}$$

which form the Lie algebra of the Lorentz group.

2.2 Two-by-two representation of the Lorentz group

It is possible to form two-by-two matrices

$$J_1 = \frac{1}{2}\sigma_1, \quad J_2 = \frac{1}{2}\sigma_2, \quad \text{and} \quad J_3 = \frac{1}{2}\sigma_3, \tag{2.21}$$

where

$$\sigma_1 = \begin{pmatrix} 0 & 1 \\ 1 & 0 \end{pmatrix}, \quad \sigma_2 = \begin{pmatrix} 0 & -i \\ i & 0 \end{pmatrix}, \quad \text{and} \quad \sigma_3 = \begin{pmatrix} 1 & 0 \\ 0 & -1 \end{pmatrix}, \tag{2.22}$$

are the Pauli spin matrices. The matrices given in equations (2.21) and (2.22) are Hermitian and their role in physics is well known.

In addition, it is also possible to form the following three anti-Hermitian matrices:

$$K_1 = \frac{i}{2}\sigma_1, \quad K_2 = \frac{i}{2}\sigma_2, \quad \text{and} \quad K_3 = \frac{i}{2}\sigma_3. \tag{2.23}$$

These matrices together with J_i satisfy the Lie algebra of the Lorentz group given in equation (2.20). They generate a group of two-by-two unimodular matrices, known as

$SL(2, c)$, where the determinant of each matrix is 1. The matrices representing $SL(2, c)$ are the smallest matrices containing as many as six degrees of freedom. For this reason, they occupy an important place in mathematics with applications in many branches of physics. For instance, these two-by-two matrices play the central role in optical sciences [4]. This aspect has been discussed extensively in the literature [1, 2, 4, 5, 11, 14, 17].

The most general form for an $SL(2, c)$ matrix can be written as

$$G = \begin{pmatrix} \alpha & \beta \\ \gamma & \delta \end{pmatrix}, \tag{2.24}$$

where all the elements are complex numbers. There are thus eight real numbers, but only six of them are independent since the determinant of this matrix is to be 1. This matrix with six independent numbers can also be expressed in the form

$$e^{-i\sum_i (\theta_i J_i + \lambda_i K_i)}, \tag{2.25}$$

with the six real parameters of θ_i and λ_i. Expanding this exponential form in a Taylor series will result in a two-by-two matrix like that of equation (2.24).

It is further remarkable that this two-by-two representation shares the same Lie algebra with the four-by-four representation for the Lorentz group. If two groups share the same Lie algebra, they are *locally isomorphic to each other* in the language of group theory, which we shall more simply say as one is *like* the other.

Indeed, as the Lorentz group is like the group $SL(2, c)$ it is said that $SL(2, c)$ is the covering group of the Lorentz group. However, we shall simply say that $SL(2, c)$ is the two-by-two representation of the Lorentz group.

For each two-by-two matrix, there is a corresponding four-by-four matrix, and it is possible to write the most general form of the four-by-four matrix in terms of the complex parameters α, β, γ and δ [11]. However, it is not necessary to use these complicated expressions here, so we list in table 2.1 only particular cases that are useful for the topics covered in this book.

2.3 Four-vectors in the two-by-two representation

The four-dimensional Minkowskian space consists of the coordinate variables t, z, and x, y, which define a vector in this four-dimensional space. It is then appropriate to call this vector a *four-vector*. Additionally, the energy–momentum four-vector can be defined as (p_0, p_z, p_x, p_y), with $p_0 = E$.

In the four-dimensional space of (t, z, x, y), we apply four-by-four matrices to perform transformations. For Lorentz transformations, the quantity

$$t^2 - z^2 - x^2 - y^2 \tag{2.26}$$

remains constant. Similarly, under Lorentz transformations, the energy–momentum four-vector

$$p_0^2 - p_z^2 - p_x^2 - p_y^2 \tag{2.27}$$

remains invariant.

Table 2.1. Generators and transformation matrices of $SL(2, c)$, and their corresponding four-by-four transformation matrices in the Lorentz group. The four-by-four matrices are applicable to the Minkowskian space of (t, z, x, y).

Generators	Two-by-two	Four-by-four
$J_1 = \frac{1}{2}\begin{pmatrix} 0 & 1 \\ 1 & 0 \end{pmatrix}$	$\begin{pmatrix} \cos(\theta/2) & i\sin(\theta/2) \\ i\sin(\theta/2) & \cos(\theta/2) \end{pmatrix}$	$\begin{pmatrix} 1 & 0 & 0 & 0 \\ 0 & \cos\theta & 0 & \sin\theta \\ 0 & 0 & 1 & 0 \\ 0 & -\sin\theta & 0 & \cos\theta \end{pmatrix}$
$K_1 = \frac{1}{2}\begin{pmatrix} 0 & i \\ i & 0 \end{pmatrix}$	$\begin{pmatrix} \cosh(\lambda/2) & \sinh(\lambda/2) \\ \sinh(\lambda/2) & \cosh(\lambda/2) \end{pmatrix}$	$\begin{pmatrix} \cosh\lambda & 0 & \sinh\lambda & 0 \\ 0 & 1 & 0 & 0 \\ \sinh\lambda & 0 & \cosh\lambda & 0 \\ 0 & 0 & 0 & 1 \end{pmatrix}$
$J_2 = \frac{1}{2}\begin{pmatrix} 0 & -i \\ i & 0 \end{pmatrix}$	$\begin{pmatrix} \cos(\theta/2) & -\sin(\theta/2) \\ \sin(\theta/2) & \cos(\theta/2) \end{pmatrix}$	$\begin{pmatrix} 1 & 0 & 0 & 0 \\ 0 & \cos\theta & -\sin\theta & 0 \\ 0 & \sin\theta & \cos\theta & 0 \\ 0 & 0 & 0 & 1 \end{pmatrix}$
$K_2 = \frac{1}{2}\begin{pmatrix} 0 & 1 \\ -1 & 0 \end{pmatrix}$	$\begin{pmatrix} \cosh(\lambda/2) & -i\sinh(\lambda/2) \\ i\sinh(\lambda/2) & \cosh(\lambda/2) \end{pmatrix}$	$\begin{pmatrix} \cosh\lambda & 0 & 0 & \sinh\lambda \\ 0 & 1 & 0 & 0 \\ 0 & 0 & 1 & 0 \\ \sinh\lambda & 0 & 0 & \cosh\lambda \end{pmatrix}$
$J_3 = \frac{1}{2}\begin{pmatrix} 1 & 0 \\ 0 & -1 \end{pmatrix}$	$\begin{pmatrix} e^{i\phi/2} & 0 \\ 0 & e^{-i\phi/2} \end{pmatrix}$	$\begin{pmatrix} 1 & 0 & 0 & 0 \\ 0 & 1 & 0 & 0 \\ 0 & 0 & \cos\phi & -\sin\phi \\ 0 & 0 & \sin\phi & \cos\phi \end{pmatrix}$
$K_3 = \frac{1}{2}\begin{pmatrix} i & 0 \\ 0 & -i \end{pmatrix}$	$\begin{pmatrix} e^{\eta/2} & 0 \\ 0 & e^{-\eta/2} \end{pmatrix}$	$\begin{pmatrix} \cosh\eta & \sinh\eta & 0 & 0 \\ \sinh\eta & \cosh\eta & 0 & 0 \\ 0 & 0 & 1 & 0 \\ 0 & 0 & 0 & 1 \end{pmatrix}$

The dot product or scalar product of two four-vectors, $p \cdot x$, is another important Lorentz-invariant quantity in physics. This can be written as

$$p \cdot x = p_0 t - p_z z - p_x x - p_y y. \qquad (2.28)$$

However, these four-vectors need to be written in the two-by-two formalism. We know there are two-by-two matrices that are applicable to two-component column vectors. They are called *spinors*. We need to understand how these four-vectors can be expressed in terms of $SL(2, c)$ spinors. For this purpose we shall start with the Hermitian matrix

$$[X] = \begin{pmatrix} t + z & x - iy \\ x + iy & t - z \end{pmatrix}. \qquad (2.29)$$

The determinant of this matrix is the Lorentz-invariant quantity given in equation (2.26). The energy–momentum four-vector can thus be written as

$$[P] = \begin{pmatrix} p_0 + p_z & p_x - ip_y \\ p_x + ip_y & p_0 - p_z \end{pmatrix}. \tag{2.30}$$

The determinant of this matrix leads to the invariant quantity of equation (2.27).
We can then write the matrix

$$[P + X] = \begin{pmatrix} p_0 + p_z + t + z & p_x - ip_y + x - iy \\ p_x + ip_y + x + iy & p_0 - p_z + t - z \end{pmatrix}. \tag{2.31}$$

The determinant of this expression gives [4, 17]

$$p \cdot x = \frac{1}{2}(\det[P + X] - \det[P] - \det[X]). \tag{2.32}$$

We can go to table 2.1, and use the two-by-two boost matrix along the z-direction and call it $B(\eta)$:

$$B(\eta) = \begin{pmatrix} e^{\eta/2} & 0 \\ 0 & e^{-\eta/2} \end{pmatrix}. \tag{2.33}$$

By taking the matrix product $B(\eta)[X]B^{\dagger}(\eta)$, we obtain

$$B(\eta)[X]B^{\dagger}(\eta) = \begin{pmatrix} (t + z)e^{\eta} & x - iy \\ x + iy & (t - z)e^{-\eta} \end{pmatrix}, \tag{2.34}$$

which results in the transformation

$$t \to (\cosh \eta)t + (\sinh \eta)z \quad \text{and} \quad z \to (\sinh \eta)t + (\cosh \eta)z, \tag{2.35}$$

which is clearly a Lorentz boost along the z-direction.
Also from table 2.1, the rotation matrix around the z-axis can be obtained and written as

$$Z(\phi) = \begin{pmatrix} e^{i\phi/2} & 0 \\ 0 & e^{-i\phi/2} \end{pmatrix}. \tag{2.36}$$

Then

$$Z(\phi)[X]Z^{\dagger}(\phi) = \begin{pmatrix} t + z & (x - iy)e^{i\phi} \\ (x + iy)e^{-i\phi} & t - z \end{pmatrix} \tag{2.37}$$

results in a rotation around the z-axis:

$$x \to (\cos \phi)x - (\sin \phi)y \quad \text{and} \quad y \to (\sin \phi)x + (\cos \phi)y. \tag{2.38}$$

Finally, we can consider a rotation around an axis perpendicular to the z-axis, and choose the rotation around the y-axis. This rotation matrix takes the form

$$R(\theta) = \begin{pmatrix} \cos(\theta/2) & -\sin(\theta/2) \\ \sin(\theta/2) & \cos(\theta/2) \end{pmatrix}. \tag{2.39}$$

If applied to the four-vector X in the same manner as in the cases of $B(\eta)$ of $Z(\phi)$, we obtain the result

$$R(\theta)[X]R^{\dagger}(\theta) = \begin{pmatrix} t + z' & x' - iy \\ x' + iy & t - z' \end{pmatrix}, \tag{2.40}$$

with

$$\begin{pmatrix} z' \\ x' \end{pmatrix} = \begin{pmatrix} \cos\theta & -\sin\theta \\ \sin\theta & \cos\theta \end{pmatrix} \begin{pmatrix} z \\ x \end{pmatrix}. \tag{2.41}$$

As the Lie algebra of the three-dimensional rotation group given in equation (2.5) indicates, rotations around two orthogonal directions can lead to a rotation around the third axis. The Lie algebra of equation (2.20) confirms that the Lorentz boost along one direction can be applied to all directions. Hence, those three transformations given in equations (2.34), (2.37), and (2.40) define the most general transformations in the four-dimensional Minkowskian space.

2.4 Transformation properties in the two-by-two representation

Transformations are straight-forward in the four-by-four representation, because four-by-four matrices are applied to the four-component vector. However, in the two-by-two representation, the four-vector is written in terms of a two-by-two matrix. Thus the momentum four-vector is, for example, written in the form given in equation (2.30).

For a particle at rest, the momentum–energy four-vector becomes

$$[P] = \begin{pmatrix} m & 0 \\ 0 & m \end{pmatrix}. \tag{2.42}$$

If this four-vector is rotated around the z- and y-directions we obtain

$$Z(\phi)[P]Z^{\dagger}(\phi) = R(\theta)[P]R^{\dagger}(\theta) = [P], \tag{2.43}$$

since rotation matrices are Hermitian. The two-by-two matrices for $Z(\phi)$ and $R(\theta)$ are given in equations (2.36) and (2.39), respectively. We can also write this expression as

$$Z(\phi)[P]Z^{-1}(\phi) = R(\theta)[P]R^{-1}(\theta) = [P], \tag{2.44}$$

because the rotation matrices are Hermitian.

As for the boost matrix of equation (2.33), its Hermitian conjugate is not its inverse. It remains invariant. Thus,

$$B(\eta)[P]B^{\dagger}(\eta) = \begin{pmatrix} e^{\eta/2} & 0 \\ 0 & e^{-\eta/2} \end{pmatrix} \begin{pmatrix} m & 0 \\ 0 & m \end{pmatrix} \begin{pmatrix} e^{\eta/2} & 0 \\ 0 & e^{-\eta/2} \end{pmatrix} = \begin{pmatrix} me^{\eta} & 0 \\ 0 & me^{-\eta} \end{pmatrix}, \qquad (2.45)$$

while $B(\eta)[P]B^{-1}(\eta) = [P]$ is

$$\begin{pmatrix} e^{\eta/2} & 0 \\ 0 & e^{-\eta/2} \end{pmatrix} \begin{pmatrix} m & 0 \\ 0 & m \end{pmatrix} \begin{pmatrix} e^{-\eta/2} & 0 \\ 0 & e^{\eta/2} \end{pmatrix} = \begin{pmatrix} m & 0 \\ 0 & m \end{pmatrix}. \qquad (2.46)$$

This difference occurs because not all the matrices of $SL(2, c)$ are Hermitian.

2.5 Subgroups of the Lorentz group

The nine commutations relations in equation (2.20) are invariant under Hermitian conjugation. Among these generators, the rotation generators J_i are invariant under Hermitian conjugation, but the boost generators K_i are anti-Hermitian with $K_i^* = -K_i$. Therefore, for every Lie algebra of equation (2.20), there is another Lie algebra where K_i is replaced by

$$\dot{K}_i = -K_i. \qquad (2.47)$$

This means that there is a dotted algebra leading to the dotted representation which corresponds to every Lie algebra of the Lorentz group. These two groups are subgroups to each other.

In this section, we are interested in subsets of those nine commutation relations which can be grouped into a closed set. The rotation generators J_i satisfy the closed set of three commutation relations. Therefore, the rotation group is a subgroup of the Lorentz group. These generators are Hermitian while the boost generators are not. Thus, the rotation subgroup is the Hermitian subgroup of the Lorentz group.

We shall only consider the two-by-two representations in the following discussion. Among the six generators, K_1, K_3, and J_2 are pure imaginary. The transformation matrices therefore have real elements and satisfy the following set of commutation relations,

$$[K_1, K_3] = iJ_2, \quad [J_2, K_3] = iK_1, \quad \text{and} \quad [J_2, K_1] = -iK_3. \qquad (2.48)$$

This subgroup of the Lorentz group consists of a rotation around the y-direction and boosts along the x- and z-directions. As in the case of the rotation subgroup, there are many physical applications for this subgroup and it plays an important role in the optical sciences. For example, this subgroup can be used in discussing the repeated application of K_3 and J_2 to the attenuation matrix or in discussing the Berry phase in polarization optics [9, 13]. Since J_2, K_1, and K_3 form a closed set of commutation relations they form the Lie algebra for the two-dimensional symplectic or $Sp(2)$ group [7] which is a three-parameter group with real elements. This three-parameter subgroup has been extensively discussed in connection with squeezed states of light [12, 16, 21].

Another interesting subgroup can be found by considering K_1 and J_3. The commutation relation of these two matrices gives

$$[J_3, K_1] = iK_2. \tag{2.49}$$

These generators form the Lie algebra for the $O(2, 1)$-like Lorentz subgroup which is applicable to two space and one time dimensions. This three-parameter subgroup has many applications in physics.

It is also possible to consider the following combinations of generators

$$J_3, \quad N_1 = K_1 - J_2 \quad \text{and} \quad N_2 = K_2 + J_1, \tag{2.50}$$

with their explicit expressions

$$J_3 = \frac{1}{2}\begin{pmatrix} 1 & 0 \\ 0 & -1 \end{pmatrix}, \quad N_1 = \begin{pmatrix} 0 & i \\ 0 & 0 \end{pmatrix}, \quad \text{and} \quad N_2 = \begin{pmatrix} 0 & 1 \\ 0 & 0 \end{pmatrix}. \tag{2.51}$$

These generators satisfy the following closed set of commutation relations:

$$[N_1, N_2] = 0, \quad [J_3, N_1] = iN_2, \quad \text{and} \quad [J_3, N_2] = -iN_1. \tag{2.52}$$

Unlike the rotation subgroup and the subgroup generated by equation (2.48), it is difficult to construct a geometry corresponding to this Lie algebra.

Wigner, however, did consider the two-dimensional Euclidean transformations consisting of one rotation and two translations along two orthogonal directions [19]. If we consider the two-dimensional space of x and y, the rotation generator is

$$J_3 = -i\left(x\frac{\partial}{\partial y} - y\frac{\partial}{\partial x} \right), \tag{2.53}$$

and the translation generators are

$$P_1 = -i\frac{\partial}{\partial x} \quad \text{and} \quad P_2 = -i\frac{\partial}{\partial y}. \tag{2.54}$$

Then the commutation relations

$$[P_1, P_2] = 0, \quad [J_3, P_1] = iP_2, \quad \text{and} \quad [J_3, P_2] = -iP_1 \tag{2.55}$$

are the same as the set given in equation (2.52). Thus the subgroup generated by the Lie algebra of equation (2.52) is like (locally isomorphic to) the two-dimensional Euclidean group [19].

2.6 Decompositions of the $Sp(2)$ matrices

The group $Sp(2)$ has many interesting decompositions which are useful in optics and other branches of physics. Among them are the Bargmann and the Iwasawa decompositions which are detailed below.

2.6.1 Bargmann decomposition

In his 1947 paper [1], Bargmann considered

$$W = \begin{pmatrix} \alpha & \beta \\ \beta^* & \alpha^* \end{pmatrix}, \tag{2.56}$$

with $\alpha\alpha^* - \beta\beta^* = 1$. There are three independent parameters. Bargmann then observed that α and β can be written as

$$\alpha = (\cosh\eta)e^{-i(\phi+\rho)} \qquad \text{and} \qquad \beta = (\sinh\eta)e^{-i(\phi-\rho)}. \tag{2.57}$$

Then W can be decomposed into

$$W(\phi, \rho, \eta) = \begin{pmatrix} e^{-i\phi} & 0 \\ 0 & e^{i\phi} \end{pmatrix} \begin{pmatrix} \cosh\eta & \sinh\eta \\ \sinh\eta & \cosh\eta \end{pmatrix} \begin{pmatrix} e^{-i\rho} & 0 \\ 0 & e^{i\rho} \end{pmatrix}. \tag{2.58}$$

In order to transform the above expression into the decomposition which will be given in equation (9.29) of chapter 9, we take the conjugate of each of the matrices with

$$C_1 = \frac{1}{\sqrt{2}} \begin{pmatrix} 1 & i \\ i & 1 \end{pmatrix}. \tag{2.59}$$

Then $C_1 W C_1^{-1}$ leads to

$$W'(\phi, \rho, \eta) = \begin{pmatrix} \cos\phi & -\sin\phi \\ \sin\phi & \cos\phi \end{pmatrix} \begin{pmatrix} \cosh\eta & \sinh\eta \\ \sinh\eta & \cosh\eta \end{pmatrix} \begin{pmatrix} \cos\rho & -\sin\rho \\ \sin\rho & \cos\rho \end{pmatrix}. \tag{2.60}$$

We have to take another conjugate with

$$C_2 = \frac{1}{\sqrt{2}} \begin{pmatrix} 1 & 1 \\ -1 & 1 \end{pmatrix}. \tag{2.61}$$

Now the conjugate $C_2 C_1 W C_1^{-1} C_2^{-1}$ becomes

$$W''(\phi, \rho, \eta) = \begin{pmatrix} \cos\phi & -\sin\phi \\ \sin\phi & \cos\phi \end{pmatrix} \begin{pmatrix} e^{\eta} & 0 \\ 0 & e^{-\eta} \end{pmatrix} \begin{pmatrix} \cos\rho & -\sin\rho \\ \sin\rho & \cos\rho \end{pmatrix}, \tag{2.62}$$

where the combined effect of $C_2 C_1$ is

$$C_2 C_1 = \frac{1}{\sqrt{2}} \begin{pmatrix} e^{i\pi/4} & e^{i\pi/4} \\ -e^{-i\pi/4} & e^{-i\pi/4} \end{pmatrix}. \tag{2.63}$$

A general $Sp(2)$ matrix can be expressed in the form

$$\begin{pmatrix} A & B \\ C & D \end{pmatrix}, \tag{2.64}$$

where all elements A, B, C, and D are real and $AD - BC = 1$. This form is related to equation (2.56) as

$$A = \alpha + \alpha^* + \beta + \beta^*,$$
$$B = -i(\alpha - \alpha^* + \beta - \beta^*),$$
$$C = -i(\alpha - \alpha^* - \beta + \beta^*),$$
$$D = \alpha + \alpha^* - \beta - \beta^*.$$

(2.65)

In terms of the parameters ρ, η, and ϕ, these real components are expressed as

$$A = (\cosh \eta)\cos(\phi + \rho) + (\sinh \eta)\cos(\phi - \rho),$$
$$B = (\cosh \eta)\sin(\phi + \rho) + (\sinh \eta)\sin(\phi - \rho),$$
$$C = (\cosh \eta)\sin(\phi + \rho) - (\sinh \eta)\sin(\phi - \rho),$$
$$D = (\cosh \eta)\cos(\phi + \rho) - (\sinh \eta)\cos(\phi - \rho).$$

(2.66)

2.6.2 Iwasawa decomposition

The rotation subgroup, among the subgroups of $SL(2, c)$, is the most used in physics. What is still largely unknown is that the subgroup consisting of real matrices plays important roles in many branches of physics. This is particularly true in particle physics and optical sciences [4]. Since the matrix in this subgroup has three degrees of freedom, it can be decomposed as in equation (2.56).

There are many interesting results which can be obtained from this form. For instance when $\phi = \rho = 0$ matrix multiplication in equation (2.60) leads to

$$\begin{pmatrix} (\cos \theta)\cosh \eta & \sinh \eta - (\sin \theta)\cosh \eta \\ \sinh \eta + (\sin \theta)\cosh \eta & (\cos \theta)\cosh \eta \end{pmatrix}.$$

(2.67)

When $\tanh \eta = \sin \theta$, this matrix becomes

$$\begin{pmatrix} 1 & 0 \\ 2 \sinh \eta & 1 \end{pmatrix}.$$

(2.68)

This form is known as the Iwasawa decomposition [10] and results in a triangular matrix that cannot be diagonalized. The physics of this triangular form is mentioned in connection with the spread of the wave packet in section 5.3 of chapter 5. In chapter 3, the role of this matrix is discussed again as the generator of gauge transformations for massless particles.

2.7 Bilinear conformal representation of the Lorentz group

The $SL(2, c)$ group has another interesting representation. We are talking about the two-by-two matrix applicable to the two-component vector such as

$$\begin{pmatrix} u' \\ v' \end{pmatrix} = \begin{pmatrix} \alpha & \beta \\ \gamma & \delta \end{pmatrix} \begin{pmatrix} u \\ v \end{pmatrix},$$

(2.69)

where all elements are complex numbers. Unlike $SU(2)$ transformations, the quantity $|u|^2 + |v|^2$ does not always stay invariant. While the linear algebra is insensitive to the normalization constant, the meaningful number for the vector (u, v) is the ratio

$$w = \frac{u}{v}. \qquad (2.70)$$

This ratio is one complex number. The issue is the transformation of this number to another complex number. The linear transformation of equation (2.69) thus takes the form

$$w' = \frac{u'}{v'} = \frac{\alpha u + \beta v}{\gamma u + \delta v} = \frac{\alpha w + \beta}{\gamma w + \delta}. \qquad (2.71)$$

This transformation is the same as the matrix transformation given in equation (2.69).

Let us consider another transformation

$$w'' = \frac{\alpha' w' + \beta'}{\gamma' w' + \delta'}. \qquad (2.72)$$

Then the result is

$$w'' = \frac{(\alpha'\alpha + \beta'\gamma)w + (\alpha'\beta + \beta'\delta)}{(\gamma'\alpha + \delta'\gamma)w + (\gamma'\beta + \delta'\delta)}. \qquad (2.73)$$

This result is equivalent to the matrix multiplication

$$\begin{pmatrix} \alpha' & \beta' \\ \gamma' & \delta' \end{pmatrix}\begin{pmatrix} \alpha & \beta \\ \gamma & \delta \end{pmatrix} = \begin{pmatrix} \alpha'\alpha + \beta'\gamma & \alpha'\beta + \beta'\delta \\ \gamma'\alpha + \delta'\gamma & \gamma'\beta + \delta'\delta \end{pmatrix}. \qquad (2.74)$$

There are many two-by-two matrices in optics, but not all of them are rotation-like matrices generated by the three Pauli spin matrices. The $SL(2, c)$ group with six independent parameters can accommodate all of them. It is interesting to note that this two-by-two representation can also be represented by the bilinear transformations introduced in this section. This bilinear technique is particularly useful in studying polarizations of light waves [9].

References

[1] Bargmann V 1947 Irreducible unitary representations of the Lorentz group *Ann. Math.* **48** 568
[2] Başkal S, Kim Y S, and Noz M E 2014 Wigner's space-time symmetries based on the two-by-two matrices of the damped harmonic oscillators and the Poincaré sphere *Symmetry* **6** 473–515
[3] Başkal S, Kim Y S, and Noz M E 2016 Entangled harmonic oscillators and space–time entanglement *Symmetry* **8** 55–80
[4] Başkal S, Kim Y S, and Noz M E 2015 *Physics of the Lorentz Group* (Bristol: IOP Publishing)

[5] Dirac P A M 1945 Application of quaternions to Lorentz transformations *Proc. R. Irish Acad. Sect.* A **50** 261–70

[6] Gilmore R 1974 *Lie Groups, Lie Algebras, and Some of Their Applications* (Mineola, NY: Dover)

[7] Guillemin V and Sternberg S 2001 *Symplectic Techniques in Physics* (Cambridge: Cambridge University Press)

[8] Han D and Kim Y S 1988 Special relativity and interferometers *Phys. Rev.* A **37** 4494–6

[9] Han D, Kim Y S, and Noz M E 1996 Polarization optics and bilinear representation of the Lorentz group *Phys. Lett.* A **219** 26–32

[10] Iwasawa K 1949 On some types of topological groups *Ann. Math.* **50** 507

[11] Kim Y S and Noz M E 1986 *Theory and Applications of the Poincaré Group* (Dordrecht: Springer)

[12] Kim Y S and Noz M E 1991 *Phase Space Picture of Quantum Mechanics: Group Theoretical Approach* (*Lecture Notes in Physics Series* vol 40) (Singapore: World Scientific)

[13] Kitano M and Yabuzaki T 1989 Observation of Lorentz-group Berry phases in polarization optics *Phys. Lett.* A **142** 321–5

[14] Naimark M A 1954 Linear representation of the Lorentz group *Usp. Mat. Nauk* **9** 19–93
Naimark M A 1957 Linear representation of the Lorentz group *Am. Math. Soc. Transl. Ser.* **2** 379458 (Engl. transl.)
Naimark M A 1964 *Linear Representations of the Lorentz Group* (*Int. Series of Monographs in Pure and Applied Mathematics* vol 63) (Oxford: Pergamon) (Engl. transl.)

[15] Peres A and Terno D R 2004 Quantum information and relativity theory *Rev. Mod. Phys.* **76** 93–123

[16] Walls D F 1983 Squeezed states of light *Nature* **306** 141–6

[17] Wightman A S 1961 *Relations de Dispersion et Particules Elementaires* ed C DeWitt, and R Omnes (New York: Wiley) pp 150–226

[18] Wigner E P 1931 *Gruppentheorie und ihre Anwendung auf die Quantenmechanik der Atomspektren* (Braunschweig: Springer)

[19] Wigner E P 1939 On unitary representations of the inhomogeneous Lorentz group *Ann. Math.* **40** 149–204

[20] Wigner E P 1959 *Group Theory: And its Application to the Quantum Mechanics of Atomic Spectra* (New York: Academic)

[21] Yuen H P 1976 Two-photon coherent states of the radiation field *Phys. Rev.* A **13** 2226–43

IOP Publishing

Mathematical Devices for Optical Sciences

Sibel Başkal, Young S Kim, and Marilyn E Noz

Chapter 3

Internal space–time symmetries

When Newton formulated his law of gravity, he wrote down a formula applicable to two point particles. It took him 20 years to prove that his formula also works for extended objects such as the Sun and Earth, as indicated in figure 3.1.

When Einstein formulated his special relativity for point particles in 1905, he used the mathematics of Lorentz transformation developed earlier by Lorentz and Poincaré. The question is what happens when those particles have space–time extensions? The hydrogen atom is a case in point. The hydrogen atom is small enough to be regarded as a point particle obeying Einstein's law of Lorentz transformations including the energy–momentum relation $E = \sqrt{p^2 + m^2}$.

Yet, the hydrogen atom has a rich internal space–time structure, rich enough to provide the foundation of quantum mechanics. Indeed, Niels Bohr was interested in why the energy levels of the hydrogen atom are discrete. His interest led to the replacement of the orbit by a standing wave.

Before and after 1927, Einstein and Bohr met occasionally to talk about physics. It is possible that they talked about how standing-waves look to moving observers. However, there are no written records. If they did not talk about this problem, it is because there were and still are no hydrogen atoms with relativistic speed. This question will be discussed in detail in appendix A.

Unlike the hydrogen atom, photons and electrons do not have space extensions. If they have, we do not talk about them. Yet, they are known to have internal angular momenta called spin. Thus, they still have internal space–time symmetries. If the particle is at rest, the particle has a rotational degree of freedom. The classical picture of this degree of freedom is the spinning top, as described in the text book by Goldstein [7].

We are now interested in moving tops in the Lorentz-covariant world with six degrees of freedom. If the momentum of the top is given, it has three rotational degrees of freedom. This can be seen clearly when the top is at rest. It was Eugene Wigner who formulated this problem by introducing the subgroups of the Lorentz

Newton's Gravity
for Point Particles

Einstein's
Lorentz Boost
for Point Particles

20 Years

Wigner | 1939

Sun
Newton's Gravity
for Extended Objects
Earth

Wigner's Little groups
for Internal
Space-time Symmetries

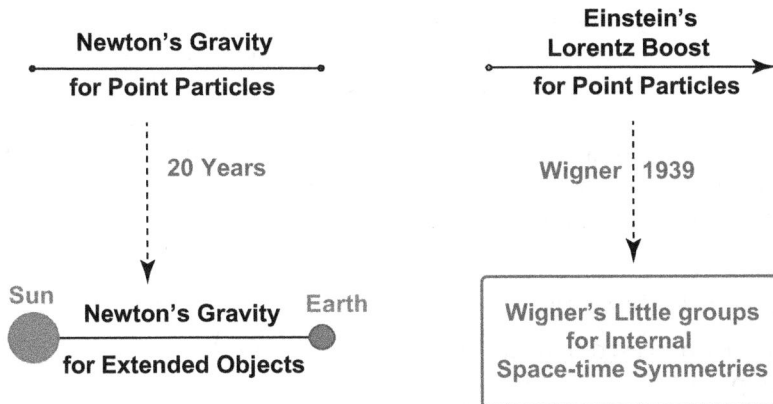

Figure 3.1. Internal space–time symmetries [15]. Einstein's special relativity for point particles is well established. What happens when those particles have internal space–time structures? What are their symmetry properties? This problem is not unlike the problem Newton had in dealing with extending his law of gravity to extended objects. It took him twenty years to formulate the new mathematics known today as *integration*. It was Eugene Wigner who, in 1939, formulated the mathematics applicable to internal space–time symmetries in the Lorentz-covariant world [20].

group whose transformations leave the momentum of a given particle invariant. Once the momentum is fixed, there are three remaining degrees of freedom. Wigner called these subgroups *little groups* in his 1939 paper [20].

In this chapter, we shall study Wigner's little groups and their transformation properties using the following three different representations:

1. The four-by-four matrices applicable to the Minkowskian space of (t, z, x, y). This four-dimensional representation allows us to construct the space–time geometry of the transformations.
2. The light-cone coordinate system with $(t + z)/\sqrt{2}$ and $(t - z)/\sqrt{2}$ instead of t and z. In this system, the four-by-four matrix becomes diagonal for Lorentz boosts along the z-direction. This light-cone system is the language of Lorentz boosts.
3. The two-by-two $SL(2, c)$ representation. Needless to say two-by-two matrices are much easier to handle than four-by-four matrices. These two-by-two matrices are also like those appearing in the literature on quantum optics, quantum information, and quantum computing. Indeed, for internal space–time symmetries, it is possible to learn lessons from the properties of the two-by-two matrices.

In section 3.1, we list three little groups as introduced by Wigner's original 1939 paper. Wigner introduced different little groups for massive, massless, and imaginary-mass particles, respectively. In section 3.2, it is noted that the light-cone coordinate system is convenient for studying Lorentz boosts, so the Lorentz-boosted little groups are studied. It is shown that all three little groups become the little group for the massless particle in the infinite-momentum limit.

In section 3.3, Wigner's little groups are formulated in the two-by-two representation of the $SL(2, c)$ group. The boost matrix is diagonal also in this representation, and boost properties are studied in detail. In section 3.4, it is shown that those three different little groups are three different branches of one expression. The continuity problem is studied in detail for the transitions from one little group to another. The mathematics of this continuity problem is studied in terms of a damped harmonic oscillator in classical mechanics in section 3.5.

3.1 Wigner's little groups

In 1939, Wigner observed that his little groups take different forms depending on the mass of the particle:

1. The particle mass is positive with $m^2 > 0$.
2. The particle is massless with $m^2 = 0$.
3. The particle mass is imaginary with $m^2 < 0$.

In his paper of 1939 [20], Wigner worked out his little groups using a specific Lorentz frame for each case. For the first case of massive particles, he worked out his little group when the particle is at rest with zero momentum. For the second case of massless particles, he assumed that the momentum is along the z-direction. For the case of imaginary-mass particles, he worked out the little group in the Lorentz frame where the energy component of the momentum–energy vanishes. We choose to call those Lorentz frames *Wigner frames* and call momentum–energy four-vectors *Wigner momentum* or *Wigner four-vector*. Those transformation matrices which leave the given four-momentum invariant will be called *Wigner matrices*.

Wigner in 1939 noted that his little group for massive particles is the three-dimensional rotation group in the Lorentz frame where the particle is at rest, as in the case of Goldstein's spinning top. Wigner noted also that his little group is like the two-dimensional Euclidean group called $E(2)$ for massless particles. We shall use the word *like* to indicate that two different groups share the same Lie algebra or the same closed set of commutation relations for their generators.

However, Wigner did not consider what happens when the massive particle is Lorentz-boosted to become light-like. The question is whether the $O(3)$-like little group becomes an $E(2)$-like little group in the limit of infinite momentum. This question has a stormy history and was not completely resolved until 1990 [17].

The classical picture of internal space–time symmetry is the rotating top, as explained in Goldstein's book on classical mechanics [7]. According to this book, the Euler angles constitute a convenient parameterization of the three-dimensional rotations. The Euler kinematics consists of two rotations around the z-axis with one rotation around the y-axis between them. These three operations cover also the rotation around the x-axis, thanks to the commutation relation

$$[J_2, J_3] = iJ_1. \tag{3.1}$$

It is thus enough to study the transformations confined to the zx-plane. The system can then be rotated around the z-axis.

What happens when this rotating top is Lorentz-boosted? The Lorentz boosts can be made along all three different directions. However, since the rotation groups can change their directions, it is sufficient to study the boosts along one of the directions, namely along the z-direction. This boost commutes with rotations around the z-axis. The question is what happens to rotations around the x- and y-directions when they are boosted along the z-direction [11]. With this point in mind, let us study the three little groups.

3.1.1 $O(3)$-like little group for massive particles

If the particle has a positive mass, there is a Lorentz frame in which it is at rest, with its four-momentum proportional to

$$P = (1, 0, 0, 0). \tag{3.2}$$

This momentum remains invariant under rotations in the Euclidean space of (z, x, y). Thus, the little group of the massive particle at rest is the three-dimensional rotation group.

The three generators of this little group are J_3, J_1, and J_2, satisfying the Lie algebra of the rotation group. The dynamical variables associated with these Hermitian operators are known to be particle spins.

This observation was made by Wigner in 1939 [20]. Wigner did not consider what happens when the system is Lorentz-boosted along the z-direction, with the four-by-four boost matrix

$$B(\eta) = \begin{pmatrix} \cosh\eta & \sinh\eta & 0 & 0 \\ \sinh\eta & \cosh\eta & 0 & 0 \\ 0 & 0 & 1 & 0 \\ 0 & 0 & 0 & 1 \end{pmatrix}. \tag{3.3}$$

If this matrix is applied to the four-momentum of equation (3.2) it becomes

$$P' = (\cosh\eta, \sinh\eta, 0, 0). \tag{3.4}$$

Let us see what happens when this same transformation is applied to the generators of the little group as

$$J'_i = B(\eta)J_i B^{-1}(\eta). \tag{3.5}$$

Under this boost operation, J_3 remains invariant, but J'_1 becomes

$$J'_1 = (\cosh\eta)J_1 + (\sinh\eta)K_2, \tag{3.6}$$

and J'_2 becomes

$$J'_2 = (\cosh\eta)J_2 - (\sinh\eta)K_1. \tag{3.7}$$

These Lorentz-boosted generators satisfy the same Lie algebra as that of the three-dimensional rotation group. Thus, the little group for the moving massive particle is still $O(3)$-like.

3.1.2 *E*(2)-like little group for massless particles

If the particle is massless, its four-momentum is proportional to

$$P = (1, 1, 0, 0). \tag{3.8}$$

This expression is of course invariant under rotations around the z-axis.

In addition, Wigner [20] observed that it is also invariant under the transformation

$$D(\gamma, \phi) = e^{-i\gamma(N_1 \cos\phi + N_2 \sin\phi)} \tag{3.9}$$

with

$$N_1 = K_1 - J_2 = \begin{pmatrix} 0 & 0 & i & 0 \\ 0 & 0 & i & 0 \\ i & -i & 0 & 0 \\ 0 & 0 & 0 & 0 \end{pmatrix}, \quad \text{and} \quad N_2 = K_2 + J_1 = \begin{pmatrix} 0 & 0 & 0 & i \\ 0 & 0 & 0 & i \\ 0 & 0 & 0 & 0 \\ i & -i & 0 & 0 \end{pmatrix}, \tag{3.10}$$

where the four-by-four matrices of J_1, J_2 and K_1, K_2 are given in chapter 2. As a consequence,

$$D(\gamma, \phi) = \begin{pmatrix} 1 + \gamma^2/2 & -\gamma^2/2 & \gamma\cos\phi & \gamma\sin\phi \\ \gamma^2/2 & 1 - \gamma^2/2 & \gamma\cos\phi & \gamma\sin\phi \\ \gamma\cos\phi & -\gamma\cos\phi & 1 & 0 \\ \gamma\sin\phi & -\gamma\sin\phi & 0 & 1 \end{pmatrix}. \tag{3.11}$$

Thus the generators of the little group are N_1, N_2, and J_3. They satisfy the following set of commutation relations

$$[N_1, N_2] = 0, \quad [N_1, J_3] = iN_2, \quad \text{and} \quad [N_2, J_3] = -iN_1. \tag{3.12}$$

As Wigner noted, this Lie algebra is the same as that for the two-dimensional Euclidean group, with

$$[P_1, P_2] = 0, \quad [P_1, J_3] = iP_2, \quad \text{and} \quad [P_2, J_3] = -iP_1, \tag{3.13}$$

where P_1 and P_2 generate translations along the x- and y-directions, respectively. They can be written as [12]

$$P_1 = -i\frac{\partial}{\partial x}, \quad \text{and} \quad P_2 = -i\frac{\partial}{\partial y}, \tag{3.14}$$

while the rotation generator J_3 takes the form

$$J_3 = -i\left(x\frac{\partial}{\partial y} - y\frac{\partial}{\partial x}\right). \tag{3.15}$$

It was noted early on, that these translational degrees correspond to gauge transformations [8, 18, 19]. Efforts have been made to explain the one-parameter gauge degree of freedom in terms of two translational degrees on the two-dimensional plane [10].

However, in 1987, Kim and Wigner considered the following operators [16]:

$$Q_1 = -ix\frac{\partial}{\partial z}, \quad \text{and} \quad Q_2 = iy\frac{\partial}{\partial z}, \tag{3.16}$$

with $x^2 + y^2 = $ constant. They generate translations along the z-direction on the surface of a circular cylinder as described in figure 3.2. Then they satisfy the following commutation relations:

$$\left[Q_1, Q_2\right] = 0, \quad \left[Q_1, J_3\right] = iQ_2, \quad \text{and} \quad \left[Q_2, J_3\right] = -iQ_1. \tag{3.17}$$

We can say that this is the Lie algebra for the *cylindrical group*. Figure 3.3 explains why the cylindrical group is isomorphic to the Euclidean group. They are both based on the surfaces tangential to a sphere.

Let us consider a photon whose momentum is along the z-direction. It has the four-potential

$$\left(A_0, A_z, A_x, A_y\right). \tag{3.18}$$

According to the Lorentz condition, $A_0 = A_z$. Thus the four-potential is

$$\left(A_0, A_0, A_x, A_y\right). \tag{3.19}$$

If we apply the $D(\gamma, \phi)$ of equation (3.11),

$$\begin{pmatrix} 1 + \gamma^2/2 & -\gamma^2/2 & \gamma\cos\phi & \gamma\sin\phi \\ \gamma^2/2 & 1 - \gamma^2/2 & \gamma\cos\phi & \gamma\sin\phi \\ \gamma\cos\phi & -\gamma\cos\phi & 1 & 0 \\ \gamma\sin\phi & -\gamma\sin\phi & 0 & 1 \end{pmatrix} \begin{pmatrix} A_0 \\ A_0 \\ A_x \\ A_y \end{pmatrix}, \tag{3.20}$$

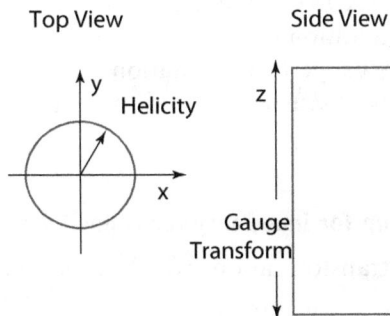

Figure 3.2. Cylindrical picture of the internal space–time structure of photons. The top view of this cylinder is a circle whose rotational degree of freedom corresponds to the helicity of the photon, while the top-down translation corresponds to a gauge transformation [17].

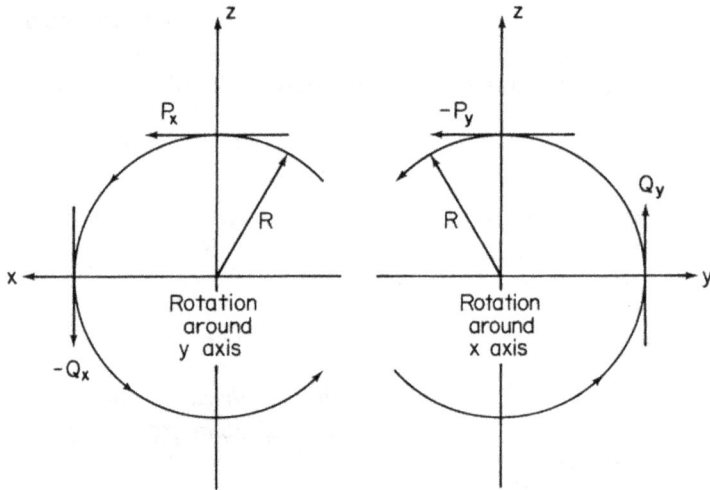

Figure 3.3. Geometry of the Euclidean and cylindrical groups. For a given sphere, there can be a plane tangential at the north pole. There can also be a cylinder tangential to the equatorial belt. The gauge transformation corresponds to the up–down translation on this cylindrical surface [16].

the result is

$$\begin{pmatrix} A_0 + \gamma\left(A_x \cos\phi + A_y \sin\phi\right) \\ A_0 + \gamma\left(A_x \cos\phi + A_y \sin\phi\right) \\ A_x \\ A_y \end{pmatrix}. \tag{3.21}$$

If we boost the four-momentum of equation (3.8) along the z-direction, the four-momentum becomes

$$P' = e^{\eta}(1, 1, 0, 0), \tag{3.22}$$

and N_1 and N_2 become $e^{\eta}N_1$ and $e^{\eta}N_2$, respectively. J_3 remains invariant.

The transformation matrix $D(\gamma, \phi)$ leaves the transverse components A_x and A_y invariant, but provides an addition to A_0. This is a cylindrical transformation. In the language of physics, it is a gauge transformation. This is summarized in table 3.1. This limiting process will be discussed in detail in section 3.2 and in more detail in section 3.4.

3.1.3 $O(2,1)$-like little group for imaginary-mass particles

We are now interested in transformations which leave the four-vector of the form

$$P = (0, 1, 0, 0) \tag{3.23}$$

invariant. Then $P^2 = -1$, and it is a negative number. We are accustomed to positive values of $P^2 = (\text{mass})^2$. This means that the particle mass is imaginary, and it moves

Table 3.1. Covariance of the energy–momentum relation, and covariance of the internal space–time symmetry. Under the Lorentz boost along the z-direction, J_3 remains invariant, and this invariant component of the angular momentum is called the helicity. The transverse components J_1 and J_2 collapse into a gauge transformation. The γ parameter for the massless case has been studied in earlier papers in the four-by-four matrix formulation [9, 11, 17].

Massive, slow	Covariance	Massless, fast
$E = p^2/2m$	Einstein's $E = mc^2$	$E = cp$
J_3	Wigner's little group	Helicity
J_1, J_2		Gauge transformation

faster than light. We realize that we are talking about a particle we cannot observe in the real world, yet its mathematics plays important roles in physics.

We are now interested in transformations which leave the four-vector of equation (3.23) invariant. Let us consider the Lorentz boost along the x-direction, with

$$S(\lambda) = \begin{pmatrix} \cosh \lambda & 0 & \sinh \lambda & 0 \\ 0 & 1 & 0 & 0 \\ \sinh \lambda & 0 & \cosh \lambda & 0 \\ 0 & 0 & 0 & 1 \end{pmatrix}, \tag{3.24}$$

which is generated by K_1. Likewise, it is invariant under the rotation around the z-axis generated by J_3. Thus, we can consider the set of commutation relations

$$[J_3, K_1] = iK_2, \quad [J_3, K_2] = -iK_1, \quad \text{and} \quad [K_1, K_2] = -iJ_3. \tag{3.25}$$

This is a Lie algebra of the Lorentz group applicable to two space dimensions and one time dimensions. This group is known in the literature as $O(2, 1)$.

If we boost the four-momentum of equation (3.23) along the z-direction using equation (3.3), it becomes

$$(\sinh \eta, \cosh \eta, 0, 0). \tag{3.26}$$

While J_3 remains invariant, K_1 and K_2 become

$$K_1' = (\cosh \eta)K_1 - (\sinh \eta)J_2, \quad \text{and} \quad K_2' = (\cosh \eta)K_2 + (\sinh \eta)J_1, \tag{3.27}$$

respectively. The generators K_1', K_2', and J_3 satisfy the same Lie algebra as that of equation (3.25). What happens when η becomes very large? This question is the same as the one we asked earlier for the $O(3)$-like little group.

Even though we are talking about imaginary-mass particles in this section, the mathematics of this little group is applicable to many branches of physics. As for the Lorentz group applicable to the $(2 + 1)$-dimensional space, physical applications seldom go beyond transformations applicable in the z- and x-coordinates. Calculations in high-energy physics involving Lorentz transformations are mostly based on this smaller group $O(2, 1)$ [13]. In addition, this group serves as one of the basic languages in classical and modern optics [2, 14], as will be discussed in later chapters of this book.

We can summarize the contents of this section with the Wigner momenta and Wigner matrices, as shown in table 3.2. Since the rotation around the z-axis commutes with the boost along the same direction, it is sufficient to consider the transformations in the zx-plane. For the most general case, the system can be rotated around the z-axis. Since this rotation commutes with the boost along the same direction, there are no complications.

3.2 Little groups in the light-cone coordinate system

In this section, we approach the same problem using Dirac's light-cone coordinate system [6], where the variables z_+ and z_- are defined as

$$z_+ = \frac{t+z}{\sqrt{2}}, \qquad \text{and} \qquad z_- = \frac{t-z}{\sqrt{2}}, \qquad (3.28)$$

and we have to work with the four-vector (z_+, z_-, x, y).

Likewise, for the four-momentum, we use

$$p_+ = \frac{p_0 + p_z}{\sqrt{2}}, \qquad \text{and} \qquad p_- = \frac{p_0 - p_z}{\sqrt{2}}. \qquad (3.29)$$

Thus, the transformation to the light-cone system is

$$\begin{pmatrix} p_+ \\ p_- \\ p_x \\ p_y \end{pmatrix} = \begin{pmatrix} 1/\sqrt{2} & 1/\sqrt{2} & 0 & 0 \\ 1/\sqrt{2} & -1/\sqrt{2} & 0 & 0 \\ 0 & 0 & 1 & 0 \\ 0 & 0 & 0 & 1 \end{pmatrix} \begin{pmatrix} p_0 \\ p_z \\ p_x \\ p_y \end{pmatrix}. \qquad (3.30)$$

Table 3.2. Wigner momenta and Wigner matrices. The matrices given in this table are transformations within the zx-plane, the y-coordinate remains invariant. However, there is one additional degree of freedom, namely the rotation around the z-axis. The Wigner momenta (p_0, p_z, p_x, p_y) remain invariant, but the matrices become rotated [5].

Mass	Wigner momentum	Wigner matrix
Massive	$\begin{pmatrix} 1 \\ 0 \\ 0 \\ 0 \end{pmatrix}$	$\begin{pmatrix} 1 & 0 & 0 & 0 \\ 0 & \cos\theta & -\sin\theta & 0 \\ 0 & \sin\theta & \cos\theta & 0 \\ 0 & 0 & 0 & 1 \end{pmatrix}$
Massless	$\begin{pmatrix} 1 \\ 1 \\ 0 \\ 0 \end{pmatrix}$	$\begin{pmatrix} 1 + \gamma^2/2 & -\gamma^2/2 & \gamma & 0 \\ \gamma^2/2 & 1 - \gamma^2/2 & \gamma & 0 \\ \gamma & -\gamma & 1 & 0 \\ 0 & 0 & 0 & 1 \end{pmatrix}$
Imaginary mass	$\begin{pmatrix} 0 \\ 1 \\ 0 \\ 0 \end{pmatrix}$	$\begin{pmatrix} \cosh\lambda & 0 & \sinh\lambda & 0 \\ 0 & 1 & 0 & 0 \\ \sinh\lambda & 0 & \cosh\lambda & 0 \\ 0 & 0 & 0 & 1 \end{pmatrix}$

The major advantage of the light-cone variables is that the matrix for the Lorentz boost is diagonal. It can be written as

$$B(\eta) = \begin{pmatrix} e^{\eta} & 0 & 0 & 0 \\ 0 & e^{-\eta} & 0 & 0 \\ 0 & 0 & 1 & 0 \\ 0 & 0 & 0 & 1 \end{pmatrix}. \tag{3.31}$$

In this light-cone coordinate system, the Wigner vector for the massive particle is proportional to $(1, 1, 0, 0)$, and the Wigner matrix takes the form

$$\begin{pmatrix} (1 + \cos\theta)/2 & (1 - \cos\theta)/2 & -\sin\theta/\sqrt{2} & 0 \\ (1 - \cos\theta)/2 & (1 + \cos\theta)/2 & \sin\theta/\sqrt{2} & 0 \\ \sin\theta/\sqrt{2} & -\sin\theta/\sqrt{2} & \cos\theta & 0 \\ 0 & 0 & 0 & 1 \end{pmatrix}. \tag{3.32}$$

For the massless particle, the Wigner momentum becomes $(1, 0, 0, 0)$ in the light-cone coordinate system. The Wigner matrix becomes

$$\begin{pmatrix} 1 & \gamma^2/2 & \gamma & 0 \\ 0 & 1 & 0 & 0 \\ 0 & \gamma & 1 & 0 \\ 0 & 0 & 0 & 1 \end{pmatrix}. \tag{3.33}$$

There is a scale change of the $\sqrt{2}$ on the γ-parameter from the expression given in equation (3.11). This change has no effect on our reasoning.

For the imaginary-mass particle, the Wigner momentum becomes $(1, -1, 0, 0)$ and the Wigner matrix is

$$\begin{pmatrix} (\cosh\lambda + 1)/2 & (\cosh\lambda - 1)/2 & \sinh\lambda/\sqrt{2} & 0 \\ (\cosh\lambda - 1)/2 & (\cosh\lambda + 1)/2 & \sinh\lambda/\sqrt{2} & 0 \\ \sinh\lambda/\sqrt{2} & \sinh\lambda/\sqrt{2} & \cosh\lambda & 0 \\ 0 & 0 & 0 & 1 \end{pmatrix}. \tag{3.34}$$

These three Wigner matrices are tabulated in table 3.3. Among the three Wigner matrices given there, the matrix for the massless particle is much simpler than the other two. We are thus interested in whether these two complicated matrices can be collapsed into the simple Wigner matrix for the massless case. We know from physics that the massive particle becomes like the massless particle. We thus Lorentz-boost those two complicated matrices using the diagonal boost matrix $B(\eta)$ given in equation (3.31).

For the massive particle, the Wigner vector becomes $(e^{\eta}, e^{-\eta}, 0, 0)$. It becomes $e^{\eta}(1, 0, 0, 0)$ in the large-η limit. The boosted Wigner matrix becomes

Table 3.3. Wigner momenta and Wigner matrices in the light-cone coordinate system. For the reasons explained in table 3.2, the momenta (p_+, p_0, p_-) and matrices are written for the three-dimensional space of t, z, and x.

Mass	Wigner momentum	Wigner matrix
Massive	$\begin{pmatrix} 1 \\ 1 \\ 0 \end{pmatrix}$	$\begin{pmatrix} (1 + \cos\theta)/2 & (1 - \cos\theta)/2 & -\sin\theta/\sqrt{2} \\ (1 - \cos\theta)/2 & (1 + \cos\theta)/2 & \sin\theta/\sqrt{2} \\ \sin\theta/\sqrt{2} & -\sin\theta/\sqrt{2} & \cos\theta \end{pmatrix}$
Massless	$\begin{pmatrix} 1 \\ 0 \\ 0 \end{pmatrix}$	$\begin{pmatrix} 1 & \gamma^2/2 & \gamma \\ 0 & 1 & 0 \\ 0 & \gamma & 1 \end{pmatrix}$
Imaginary mass	$\begin{pmatrix} 1 \\ -1 \\ 0 \end{pmatrix}$	$\begin{pmatrix} (\cosh\lambda + 1)/2 & (\cosh\lambda - 1)/2 & \sinh\lambda/\sqrt{2} \\ (\cosh\lambda - 1)/2 & (\cosh\lambda + 1)/2 & \sinh\lambda/\sqrt{2} \\ \sinh\lambda/\sqrt{2} & \sinh\lambda/\sqrt{2} & \cosh\lambda \end{pmatrix}$

$$\begin{pmatrix} (1 + \cos\theta)/2 & e^{2\eta}(1 - \cos\theta)/2 & -e^{\eta}\sin\theta/\sqrt{2} & 0 \\ e^{-2\eta}(1 - \cos\theta)/2 & (1 + \cos\theta)/2 & e^{-\eta}\sin\theta/\sqrt{2} & 0 \\ e^{-\eta}\sin\theta/\sqrt{2} & -e^{\eta}\sin\theta/\sqrt{2} & \cos\theta & 0 \\ 0 & 0 & 0 & 1 \end{pmatrix}. \tag{3.35}$$

If this matrix is to remain finite, the angle θ has to be small, and $e^{\eta}\theta$ can remain finite. If we let

$$e^{\eta}\theta = -\sqrt{2}\,\gamma, \tag{3.36}$$

this boosted matrix becomes the Wigner matrix of equation (3.33).

This limiting process is illustrated by the mass hyperbola in figure 3.4. After this limit, the Wigner vector can come back to the finite-energy state through the light-cone track, also shown in figure 3.4.

For the imaginary-mass particle, the boosted Wigner matrix takes the form

$$\begin{pmatrix} (\cosh\lambda + 1)/2 & e^{2\eta}(\cosh\lambda - 1)/2 & e^{\eta}\sinh\lambda/\sqrt{2} & 0 \\ e^{-2\eta}(\cosh\lambda - 1)/2 & (\cosh\lambda + 1)/2 & e^{-\eta}\sinh\lambda/\sqrt{2} & 0 \\ e^{-\eta}\sinh\lambda/\sqrt{2} & e^{\eta}\sinh\lambda/\sqrt{2} & \cosh\lambda & 0 \\ 0 & 0 & 0 & 1 \end{pmatrix}. \tag{3.37}$$

Here again, the parameter λ should become small when η becomes very large. If we let

$$e^{\eta}\lambda = \sqrt{2}\,\gamma, \tag{3.38}$$

this boosted matrix becomes the Wigner matrix for the massless particle.

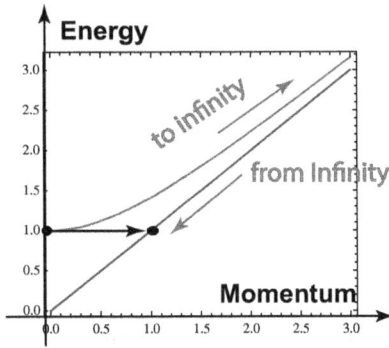

Figure 3.4. Wigner excursion. It is possible to obtain the Wigner matrix for the massless case by making exclusions specified in this figure. We can boost the Wigner matrix through the hyperbola for the massive particle to the infinite momentum where it coincides with the light cone. We can then come back through the light-cone line for the massless particle [4].

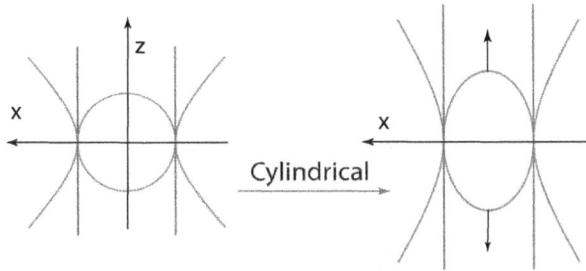

Figure 3.5. Wigner matrices for massive, massless, and imaginary-mass particles. They are illustrated with a sphere, cylinder, and the hyperbolic surface in this figure, respectively. In the high-speed limit, all three of them become cylindrical.

Figure 3.5 illustrates the collapse of these two complicated Wigner matrices into one simple matrix for the massless case. The $O(3)$-like little group for the massive particle is shown as the circle in that figure. The little group for imaginary-mass particles is illustrated as hyperbolic curves. Both the circle and hyperbola collapse into two vertical lines corresponding to the cylindrical symmetry of the $E(2)$-like little group for the massless particle.

As we noted before, the Wigner matrix for the massless case is much simpler than the other two. Let us go back to equation (3.33). If we allow the rotation around the z-axis, the matrix becomes

$$D(\gamma, \phi) = \begin{pmatrix} 1 & \gamma^2/2 & \gamma \cos\theta & \gamma \sin\theta \\ 0 & 1 & 0 & 0 \\ 0 & \gamma \cos\theta & 1 & 0 \\ 0 & \gamma \sin\theta & 0 & 1 \end{pmatrix}. \tag{3.39}$$

This is the $D(\gamma, \phi)$ matrix given in equation (3.11) written in the light-cone coordinate system. There is a scale change $\sqrt{2}$ for γ. This generates gauge transformations when applied to the photon four-vector.

3-12

Let us next consider the four-vectors, such as the electromagnetic four-potential. In the light-cone coordinate system, the four components of the vector particle take the form

$$(A_+, A_-, A_x, A_y), \tag{3.40}$$

where

$$A_+ = \frac{A_0 + A_z}{\sqrt{2}}, \qquad \text{and} \qquad A_- = \frac{A_0 - A_z}{\sqrt{2}}. \tag{3.41}$$

If it is boosted along the hyperbolic route in figure 3.4, this four-vector becomes

$$(e^\eta A_+, 0, A_x, A_y) \tag{3.42}$$

for large values of η.

If the system is boosted back along the light cone, the four-vector becomes

$$(A_+, 0, A_x, A_y). \tag{3.43}$$

During this process, A_0 vanishes. This leads to $A_0 = A_z$. The excursion given in figure 3.4 is equivalent to imposing the Lorentz boost. As for the Wigner matrix, the η dependence is gone, and it remains invariant during this return trip.

When the D matrix is applied to this photon four-vector,

$$\begin{pmatrix} 1 & \gamma^2/2 & \gamma\cos\phi & \gamma\sin\phi \\ 0 & 1 & 0 & 0 \\ 0 & -\gamma\cos\phi & 1 & 0 \\ 0 & -\gamma\sin\phi & 0 & 1 \end{pmatrix} \begin{pmatrix} A_+ \\ 0 \\ A_x \\ A_y \end{pmatrix} = \begin{pmatrix} A_+ + \gamma(A_x\cos\phi + A_y\sin\phi) \\ 0 \\ A_x \\ A_y \end{pmatrix}, \tag{3.44}$$

the transverse components A_x and A_y are not affected. The transformation changes only the A_0 and A_z components. This is a gauge transformation.

3.3 Two-by-two representation of the little groups

The Lorentz group and its two-by-two representation are discussed in chapter 2. The transformation matrix can be written as an unimodular two-by-two matrix of the form

$$G = \begin{pmatrix} \alpha & \beta \\ \gamma & \delta \end{pmatrix}, \tag{3.45}$$

with unit determinant. While all the elements are complex numbers, this matrix has six independent parameters.

The momentum four-vector can be written as

$$\begin{pmatrix} p_0 + p_z & p_x - ip_y \\ p_x + ip_y & p_0 - p_z \end{pmatrix}. \tag{3.46}$$

This four-momentum is Lorentz transformed according to

$$\begin{pmatrix} p_0' + p_z' & p_x' - ip_y' \\ p_x' + ip_y' & p_0' - p_z' \end{pmatrix} = \begin{pmatrix} \alpha & \beta \\ \gamma & \delta \end{pmatrix} \begin{pmatrix} p_0 + p_z & p_x - ip_y \\ p_x + ip_y & p_0 - p_z \end{pmatrix} \begin{pmatrix} \alpha & \beta \\ \gamma & \delta \end{pmatrix}^\dagger. \tag{3.47}$$

The Wigner momenta are proportional to

$$\begin{pmatrix} 1 & 0 \\ 0 & 1 \end{pmatrix}, \quad \begin{pmatrix} 1 & 0 \\ 0 & 0 \end{pmatrix}, \quad \text{and} \quad \begin{pmatrix} 1 & 0 \\ 0 & -1 \end{pmatrix}, \tag{3.48}$$

respectively, for the massive, massless, and imaginary-mass particles. These momentum matrices are invariant under the rotation around the z-axis with the rotation matrix

$$\begin{pmatrix} e^{i\phi/2} & 0 \\ 0 & e^{-i\phi/2} \end{pmatrix}. \tag{3.49}$$

In addition, they are invariant under transformations by the Wigner matrices given in table 3.4.

The two-by-two matrix for the Lorentz boost along the z-direction is

$$\begin{pmatrix} e^{\eta/2} & 0 \\ 0 & e^{-\eta/2} \end{pmatrix}. \tag{3.50}$$

As a consequence, the boosted Wigner vector and boosted Wigner matrix become

$$\begin{pmatrix} e^\eta & 0 \\ 0 & e^{-\eta} \end{pmatrix} \quad \text{and} \quad \begin{pmatrix} \cos(\theta/2) & -e^\eta \sin(\theta/2) \\ e^{-\eta} \sin(\theta/2) & \cos(\theta/2) \end{pmatrix}, \tag{3.51}$$

respectively. If we go through the Wigner excursion as described in figure 3.4, the result is the triangular matrix for the massless particle. The Wigner excursion for the imaginary-mass particle will lead to the same result.

Table 3.4. Wigner momenta and Wigner matrices in the two-by-two representation of the Lorentz group. This table is the two-by-two version of tables 3.2 and 3.3. The most interesting expression is the triangular matrix for the massless case.

Mass	Wigner momentum	Wigner matrix
Massive	$\begin{pmatrix} 1 & 0 \\ 0 & 1 \end{pmatrix}$	$\begin{pmatrix} \cos(\theta/2) & -\sin(\theta/2) \\ \sin(\theta/2) & \cos(\theta/2) \end{pmatrix}$
Massless	$\begin{pmatrix} 1 & 0 \\ 0 & 0 \end{pmatrix}$	$\begin{pmatrix} 1 & -\gamma \\ 0 & 1 \end{pmatrix}$
Imaginary mass	$\begin{pmatrix} 1 & 0 \\ 0 & -1 \end{pmatrix}$	$\begin{pmatrix} \cosh(\lambda/2) & \sinh(\lambda/2) \\ \sinh(\lambda/2) & \cosh(\lambda 2) \end{pmatrix}$

As for the massless case, let us consider the electromagnetic potential whose four-vector can be written as

$$\begin{pmatrix} A_0 + A_z & A_x - iA_y \\ A_x + iA_y & A_0 - A_z \end{pmatrix}. \tag{3.52}$$

After the Wigner excursion, this matrix becomes

$$\begin{pmatrix} 2A_0 & A_x - iA_y \\ A_x + iA_y & 0 \end{pmatrix}. \tag{3.53}$$

Let us go back to the triangular Wigner matrix. This matrix can be rotated around the z-direction with the rotation matrix of equation (3.49), and the result is

$$D(\gamma, \phi) = \begin{pmatrix} 1 & -e^{i\phi}\gamma \\ 0 & 1 \end{pmatrix}. \tag{3.54}$$

This expression is the two-by-two version of the $D(\gamma, \phi)$ four-by-four Wigner matrix of equation (3.11). Let us apply this two-by-two matrix to the four-vector of equation (3.52):

$$\begin{pmatrix} 1 & -e^{i\phi}\gamma \\ 0 & 1 \end{pmatrix} \begin{pmatrix} 2A_0 & A_x - iA_y \\ A_x + iA_y & 0 \end{pmatrix} \begin{pmatrix} 1 & 0 \\ -e^{-i\phi}\gamma & 1 \end{pmatrix}, \tag{3.55}$$

leading to

$$\begin{pmatrix} 2A_0 - 2\gamma(A_x \cos\phi + A_y \sin\phi) & A_x - iA_y \\ A_x + iA_y & 0 \end{pmatrix}. \tag{3.56}$$

This result tells us that the transverse components A_x and A_y remain invariant, while there is an addition of $\gamma(A_x \cos\phi + A_y \sin\phi)$ to A_0 and to A_z. As we noted in section 3.2, this is the gauge transformation.

3.4 One expression with three branches

Let us go back to equation (3.51) for the Lorentz-boosted Wigner vector for the massive particle. It is proportional to the form

$$\begin{pmatrix} 1 & 0 \\ 0 & e^{-2\eta} \end{pmatrix}. \tag{3.57}$$

The boosted Wigner matrix for the imaginary-mass particle is proportional to

$$\begin{pmatrix} 1 & 0 \\ 0 & -e^{-2\eta} \end{pmatrix}. \tag{3.58}$$

In the large η limit, these two expressions become

$$\begin{pmatrix} 1 & 0 \\ 0 & 0 \end{pmatrix}, \tag{3.59}$$

which is the Wigner vector for the massless particle. It is because

$$e^{-\eta} = -e^{-\eta} \tag{3.60}$$

for large values of η. It is continuous, but is it an analytic continuation?

In order to study this continuity of the Wigner matrices, let us consider the exponential form [1, 2, 4]

$$D(x, y) = \exp\left\{ \begin{pmatrix} 0 & -(x+y) \\ -x+y & 0 \end{pmatrix} \right\}. \tag{3.61}$$

1. If $y > x$, we write

$$x + y = e^{\eta}\sqrt{y^2 - x^2}, \qquad \text{and} \qquad y - x = e^{-\eta}\sqrt{y^2 - x^2}, \tag{3.62}$$

with

$$e^{\eta} = \sqrt{\frac{x+y}{|y-x|}}, \tag{3.63}$$

and $D(x, y)$ becomes

$$\exp\left\{ \sqrt{y^2 - x^2} \begin{pmatrix} 0 & -e^{\eta} \\ e^{-\eta} & 0 \end{pmatrix} \right\} = \begin{pmatrix} \cos(\theta/2) & -e^{\eta}\sin(\theta/2) \\ e^{-\eta}\sin(\theta/2) & \cos(\theta/2) \end{pmatrix}, \tag{3.64}$$

with $\theta/2 = \sqrt{y^2 - x^2}$. This expression is given in section 3.3 as the Wigner matrix for the massive particle.

2. If $x = y$, this expression becomes

$$D(x, y) = \exp\left\{ \begin{pmatrix} 0 & -2x \\ 0 & 0 \end{pmatrix} \right\} = \begin{pmatrix} 1 & -2x \\ 0 & 1 \end{pmatrix}. \tag{3.65}$$

This triangular form serves as the boosted Wigner matrix for massless particles, as shown in section 3.3.

3. If $x > y$, we can write

$$x - y = e^{-\eta}\sqrt{x^2 - y^2}, \qquad \text{and} \qquad x + y = e^{\eta}\sqrt{x^2 - y^2}. \tag{3.66}$$

Then the D matrix becomes

$$\exp\left\{ \sqrt{x^2 - y^2} \begin{pmatrix} 0 & -e^{\eta} \\ -e^{-\eta} & 0 \end{pmatrix} \right\} = \begin{pmatrix} \cosh(\lambda/2) & -e^{\eta}\sinh(\lambda/2) \\ -e^{-\eta}\sinh(\lambda/2) & \cosh(\lambda/2) \end{pmatrix}, \tag{3.67}$$

with

$$\lambda/2 = \sqrt{x^2 - y^2}. \tag{3.68}$$

This expression is the same as the boosted Wigner matrix for the imaginary-mass particle as shown in section 3.3.

Indeed, it is possible to derive three different forms of the Wigner matrix from one exponential form given in equation (3.61). The matrix

$$\begin{pmatrix} 0 & -(x+y) \\ -(x-y) & 0 \end{pmatrix} \tag{3.69}$$

is analytic in the x- and y-variables. However, this D matrix has three distinct branches. Let us look at what happens when $x - y$ is a small number

$$\epsilon = x - y. \tag{3.70}$$

We can then write the D matrix as

$$D(x, \epsilon) = \exp\left\{\begin{pmatrix} 0 & -2x \\ \epsilon & 0 \end{pmatrix}\right\}. \tag{3.71}$$

If ϵ is positive, the Taylor expansion leads to

$$D = \begin{pmatrix} \cosh(\sqrt{2x\epsilon}) & -[\sqrt{2x/\epsilon}]\sinh\sqrt{2x\epsilon} \\ -[\sqrt{\epsilon/2x}]\sinh(\sqrt{2x\epsilon}) & \cosh(\sqrt{2x\epsilon}) \end{pmatrix}. \tag{3.72}$$

If ϵ becomes zero, this expression becomes

$$\begin{pmatrix} 1 & -2x \\ 0 & 1 \end{pmatrix}. \tag{3.73}$$

If ϵ becomes negative,

$$\sqrt{2x\epsilon} = i\sqrt{-2x\epsilon}, \quad \sqrt{\epsilon/2x} = i\sqrt{-2x/\epsilon}, \quad \text{and} \quad \sqrt{2x/\epsilon} = -i\sqrt{-2x/\epsilon}, \tag{3.74}$$

if we take $\sqrt{-1} = i$. Thus, D becomes

$$D = \begin{pmatrix} \cos(\sqrt{-2x\epsilon}) & -[\sqrt{-2x/\epsilon}]\sin(\sqrt{-2x\epsilon}) \\ [\sqrt{-\epsilon/2x}]\sin(\sqrt{-2x\epsilon}) & \cos(\sqrt{-2x\epsilon}) \end{pmatrix}. \tag{3.75}$$

The result remains the same if we take $\sqrt{-1} = -i$.

This type of singularity is not common in the literature. We shall study this problem in section 3.5 using a physical problem very familiar to us.

3.5 Classical damped oscillators

Let us start with the second-order differential equation

$$\frac{d^2y}{dt^2} + 2\mu\frac{dy}{dt} + \omega^2 y = 0, \tag{3.76}$$

for a classical damped harmonic oscillator. If we introduce the function $\psi(t)$ as

$$\psi(t) = e^{-\mu t}y(t) \tag{3.77}$$

then $\psi(t)$ satisfies the simplified differential equation

$$\frac{d^2\psi(t)}{dt^2} + (\omega^2 - \mu^2)\psi(t) = 0. \tag{3.78}$$

This second-order differential equation has two independent solutions. Let us call them ψ_1 and ψ_2. They satisfy the first-order differential equations

$$\frac{d}{dt}\begin{pmatrix}\psi_1\\\psi_2\end{pmatrix} = \begin{pmatrix}0 & -(\omega+\mu)\\(\omega-\mu) & 0\end{pmatrix}\begin{pmatrix}\psi_1\\\psi_2\end{pmatrix}. \tag{3.79}$$

This coupled equation leads to the second-order equation (3.78) for $\psi_1(t)$ and $\psi_2(t)$. The physical solution is an appropriate linear combination of these two wave functions.

The solution of this first-order differential equation is

$$\begin{pmatrix}\psi_1\\\psi_2\end{pmatrix} = \exp\left\{\begin{pmatrix}0 & -(\omega+\mu)t\\(\omega-\mu)t & 0\end{pmatrix}\right\}\begin{pmatrix}C_1\\C_2\end{pmatrix}, \tag{3.80}$$

where $C_1 = \psi_1(0)$ and $C_2 = \psi_2(0)$. We can then obtain the solutions by following the procedure developed in section 3.4.

1. If $\omega > \mu$, the solution becomes

$$\begin{pmatrix}\psi_1\\\psi_2\end{pmatrix} = \begin{pmatrix}\cos(\omega't) & -e^\eta\sin(\omega't)\\e^{-\eta}\sin(\omega't) & \cos(\omega't)\end{pmatrix}\begin{pmatrix}C_1\\C_2\end{pmatrix}, \tag{3.81}$$

with

$$\omega' = \sqrt{\omega^2 - \mu^2}, \quad \text{and} \quad e^\eta = \sqrt{\frac{\omega+\mu}{|\omega-\mu|}}. \tag{3.82}$$

2. If $\omega = \mu$, the solution becomes

$$\begin{pmatrix}\psi_1\\\psi_2\end{pmatrix} = \begin{pmatrix}1 & -2\omega t\\0 & 1\end{pmatrix}\begin{pmatrix}C_1\\C_2\end{pmatrix}. \tag{3.83}$$

3. If $\mu > \omega$, the solution matrix becomes

$$\begin{pmatrix} \cosh(\mu't) & e^{\eta}\sinh(\mu't) \\ e^{-\eta}\sinh(\mu't) & \cosh(\mu't) \end{pmatrix}, \tag{3.84}$$

with e^{η} given in equation (3.82), and

$$\mu' = \sqrt{\mu^2 - \omega^2}. \tag{3.85}$$

Let us now turn to the main issue of what happened when μ is close to ω. If ω is sufficiently close to μ, we can let

$$\mu - \omega = 2\mu\epsilon, \quad \text{and} \quad \mu + \omega = 2\omega. \tag{3.86}$$

If ω is greater than μ, then ϵ defined in equation (3.86) becomes negative. The solution matrix becomes

$$\begin{pmatrix} 1 - (-\epsilon/2)(2\omega t)^2 & -2\omega t \\ -(-\epsilon)(2\omega t) & 1 - (-\epsilon/2)(2\omega t)^2 \end{pmatrix}, \tag{3.87}$$

which can be written as

$$\begin{pmatrix} 1 - (1/2)[(2\omega\sqrt{-\epsilon})t]^2 & -2\omega t \\ -\sqrt{-\epsilon}[(2\omega\sqrt{-\epsilon})t] & 1 - (1/2)[(2\omega\sqrt{-\epsilon})t]^2 \end{pmatrix}. \tag{3.88}$$

If ϵ is positive, equation (3.84) can be written as

$$\begin{pmatrix} 1 + (1/2)[(2\omega\sqrt{\epsilon})t]^2 & -2\omega t \\ -\sqrt{\epsilon}[(2\omega\sqrt{\epsilon})t] & 1 + (1/2)[(2\omega\sqrt{\epsilon})t]^2 \end{pmatrix}. \tag{3.89}$$

The transition from equation (3.88) to equation (3.89) is continuous as they become identical when $\epsilon = 0$. As ϵ changes its sign, the diagonal elements of the above matrices tell us how $\cos(\omega't)$ becomes $\cosh(\mu't)$. The upper-right element remains as $\sin(\omega't)$ during this transitional process. The lower left element becomes $\sinh(\mu't)$. This non-analytic continuity is illustrated in figure 3.6.

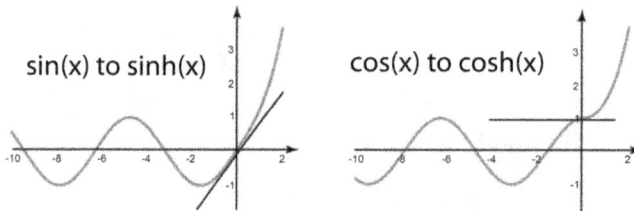

Figure 3.6. Transitions from sine to sinh, and from cosine to cosh. They are continuous transitions. Their first derivatives are also continuous, but the second derivatives are not. Thus, they are not analytically but only tangentially continuous [3].

Table 3.5. Damped oscillators and space–time symmetries. They are based on the same set of two-by-two matrices.

Trace	Damped oscillator	Particle symmetry
Smaller than 2	Oscillation mode	Massive particles
Equal to 2	Transition mode	Massless particles
Larger than 2	Damping mode	Imaginary-mass particles

During this continuation process, the function remains the same. So does its first derivative, but the second derivative does not. Thus, the two functions share the same tangential line. It is indeed a *tangential continuity*. The continuity from one little group to another was discussed in section 3.4. This mathematical similarity is summarized in table 3.5.

References

[1] Başkal S and Kim Y S 2010 One analytic form for four branches of the *ABCD* matrix *J. Mod. Opt.* **57** 1251–9
[2] Başkal S and Kim Y S 2013 Lorentz group in ray and polarization optics *Mathematical Optics: Classical, Quantum and Computational Methods* ed V Lakshminarayanan, M L Calvo, and T Alieva (Boca Raton, FL: Taylor and Francis) pp 303–49
[3] Başkal S and Kim Y S 2014 Lens optics and the continuity problems of the *ABCD* matrix *J. Mod. Opt.* **61** 161–6
[4] Başkal S, Kim Y S, and Noz M E 2014 Wigner's space–time symmetries based on the two-by-two matrices of the damped harmonic oscillators and the Poincaré sphere *Symmetry* **6** 473–515
[5] Başkal S, Kim Y S, and Noz M E 2015 *Physics of the Lorentz Group* (Bristol: IOP Publishing)
[6] Dirac P A M 1949 Forms of relativistic dynamics *Rev. Mod. Phys.* **21** 392–9
[7] Goldstein H 1980 *Classical Mechanics* (*Addison-Wesley Series in Physics*) 2nd edn (Reading, MA: Addison-Wesley)
[8] Han D and Kim Y S 1981 Little group for photons and gauge transformations *Am. J. Phys* **49** 348–51
[9] Han D, Kim Y S, and Son D 1982 *E*(2)-like little group for massless particles and neutrino polarization as a consequence of gauge invariance *Phys. Rev.* D **26** 3717–25
[10] Han D, Kim Y S, and Son D 1982 Photon spin as a rotation in gauge space *Phys. Rev.* D **25** 461–3
[11] Han D, Kim Y S, and Son D 1986 Eulerian parametrization of Wigner's little groups and gauge transformations in terms of rotations in two-component spinors *J. Math. Phys.* **27** 2228–35
[12] Inönü E and Wigner E P 1953 On the contraction of groups and their representations *Proc. Natl Acad. Sci.* **39** 510–24
[13] Kim Y S and Noz M E 1986 *Theory and Applications of the Poincaré Group* (Dordrecht: Springer)

[14] Kim Y S and Noz M E 1991 *Phase Space Picture of Quantum Mechanics: Group Theoretical Approach* (*Lecture Notes in Physics* vol 40) (Singapore: World Scientific)

[15] Kim Y S and Noz M E 2018 *New Perspectives on Einstein's* $E = mc^2$ (Singapore: World Scientific)

[16] Kim Y S and Wigner E P 1987 Cylindrical group and massless particles *J. Math. Phys.* **28** 1175–9

[17] Kim Y S and Wigner E P 1990 Space–time geometry of relativistic particles *J. Math. Phys.* **31** 55–60

[18] Kupersztych J 1976 Is there a link between gauge invariance, relativistic invariance and electron spin? *Nuovo Cim.* **B 11** 1–11

[19] Weinberg S 1964 Photons and gravitons in *S*-matrix theory: derivation of charge conservation and equality of gravitational and inertial mass *Phys. Rev.* B **135** B1049–56

[20] Wigner E P 1939 On unitary representations of the inhomogeneous Lorentz group *Ann. Math.* **40** 149–204

IOP Publishing

Mathematical Devices for Optical Sciences

Sibel Başkal, Young S Kim, and Marilyn E Noz

Chapter 4

Photons and neutrinos in the relativistic world of Maxwell and Wigner

Einstein's photo-electric effect allows us to regard electromagnetic waves as massless particles, commonly called photons. In the quantum world, the photon can have an angular momentum of 1 or −1, known as helicity, which is aligned along the direction of the momentum. Then, how is the photon helicity translated into the electric and magnetic fields perpendicular to the direction of propagation? Classically, there are electric and magnetic fields perpendicular to the direction of propagation. This aspect is translated into the polarization of photons. The question is how? This is an issue of the internal space–time symmetries defined by Wigner's little groups [33] for massive and massless particles.

In the Maxwell picture, the electromagnetic four-potential becomes a four-vector, while the electric and magnetic fields are grouped into the four-by-four Maxwell tensor. These are represented by

$$A_\mu = \begin{pmatrix} A_0 \\ A_z \\ A_x \\ A_y \end{pmatrix} \quad \text{and} \quad (F_{\mu\nu}) = \begin{pmatrix} 0 & -E_z & -E_x & -E_y \\ E_z & 0 & -B_y & B_x \\ E_x & B_y & 0 & -B_z \\ E_y & -B_x & B_z & 0 \end{pmatrix}. \tag{4.1}$$

These expressions belong to Einstein's Lorentz-covariant world. However, the issue of the electromagnetic four-potential with the gauge degree of freedom was much debated in the literature, but was settled in later papers [24, 25, 31]. As for the Maxwell tensor, the present authors dealt with the problem in their recent publications [4, 5].

In section 3.1 we saw that there are three generators for the rotation group defining the spin of a massive particle at rest. As was noted in section 1.4.1, the closed set of commutation relations defining the Lie algebra of the rotation group is a direct consequence of Heisenberg's uncertainty relation.

In section 2.2 we saw that the rotation group can be generated by three two-by-two Pauli matrices for spin-half particles known as spinors. This group of two-by-two matrices, commonly called $SU(2)$, represents two-component spinors. The direct product of two spinors yields four states leading to one spin-0 state and three spin-1 states. For the spinors we use

$$u = \begin{pmatrix} 1 \\ 0 \end{pmatrix} \quad \text{and} \quad v = \begin{pmatrix} 0 \\ 1 \end{pmatrix}, \tag{4.2}$$

for the spin-up and spin-down states, respectively.

With these spinors, we can construct the spin-0 and spin-1 states in the following manner. For the spin-0 state, we make the anti-symmetric combination $\frac{1}{\sqrt{2}}(uv - vu)$. For three spin-1 states we need symmetric combinations as uu, $\frac{1}{\sqrt{2}}(uv + vu)$, vv, for the z-component spin 1, 0, and -1, respectively.

As we also saw in section 2.2, when we add the Lorentz boost to the rotation group, it is extended to the group of two-by-two matrices called $SL(2, c)$, which serves as the covering group for the group of Lorentz transformations. From the mathematical point of view, this chapter is about the extension of the group $SU(2)$ to $SL(2, c)$ within the world of two-by-two matrices. From the physical point of view, we study here the issue of building a bridge between Heisenberg's uncertainty relation and Maxwell's Lorentz-covariant electromagnetic fields.

In discussing the subgroups of the Lorentz group in section 2.5, we noted that there is both a dotted and an undotted representation. Thus in this Lorentz-covariant world, there are two additional spin states. Therefore, in the $SL(2, c)$ regime, there are four spin states for each particle, as in the case of the Dirac equation. The direct product of two $SL(2, c)$ spinors leads to sixteen states. Among them, four can be used for the electromagnetic four-potential, and six for the Maxwell tensor. In this chapter, we present further details regarding the Maxwell four-potential and the Maxwell tensor starting from the transformation properties of the four-spinors defined in the Lorentz-covariant world.

In section 4.1 we further discuss the two-by-two representation of the Lorentz group, and introduce more concepts related to Wigner's little groups. In section 4.2 we discuss massless particles as the large-momentum or small-mass limit of massive particles. This leads to the four spin states in the Lorentz-covariant world. We show in section 4.3 how this leads to the polarization of neutrinos. In section 4.4 we construct explicitly the sixteen states resulting from the four spin states. Among them are the electromagnetic four-vector and the Maxwell tensor. This leads to the polarizarion of photons.

4.1 The Lorentz group and Wigner's little groups

As we saw in section 2.2, the rotation generators J_i and boost generators K_i can be written in two-by-two matrix format as in equations (2.21) and (2.23). When the boost generators K_i are added to the generators J_i for the rotation group, they form the Lie algebra for the $SL(2, c)$ group. In equation (2.20), the commutation relations forming this closed set are given as

$$\left[J_i, J_j\right] = i\epsilon_{ijk}J_k,$$
$$\left[J_i, K_j\right] = i\epsilon_{ijk}K_k, \quad \text{and} \qquad (4.3)$$
$$\left[K_i, K_j\right] = -i\epsilon_{ijk}K_k.$$

These two-by-two matrices perform Lorentz transformations on the four-dimensional Minkowskian space leaving invariant the quantity $(t^2 - z^2 - x^2 - y^2)$. As there are three rotation and three boost generators, the Lorentz group is a six-parameter group.

Although here we have represented the Lorentz group by two-by-two matrices, in chapter 2 we showed that for each two-by-two transformation matrix, there is a corresponding four-by-four matrix applicable to the Minkowskian space. This is illustrated in table 2.1 for some special cases needed here.

While the J_i generators are Hermitian, the K_i are not. However, the algebra of equation (4.3) is invariant under the sign change of the K_i matrices. Let us introduce the notation

$$\dot{K}_i = -K_i. \qquad (4.4)$$

Then, as was noted in section 2.5, there is another Lorentz group consisting of

$$J_i = \frac{1}{2}\sigma_i \quad \text{and} \quad \dot{K}_i = -\frac{i}{2}\sigma_i. \qquad (4.5)$$

Corresponding to these two-by-two matrices, we can construct one set of two-component spinors (spin-up and spin-down) for the undotted representation and another set for the dotted representation. There are thus four spin states in the Lorentz-covariant world as shown in table 4.1. This is the reason why the Dirac spinor has four components [21]. If four-spinors (two dotted and two undotted) are coupled, there are sixteen (4×4) states, which can be partitioned into to the spin-0 and spin-1 states. We shall come back to this problem in section 4.4.

From chapter 2, we know that the six generators of equation (4.3) lead to the group of two-by-two unimodular matrices of the form given in equation (2.24) where the matrix elements are complex numbers. Since the $\det(G) = 1$, there are six independent real numbers to accommodate the six generators J_i and K_i given in

Table 4.1. Spinors in the relativistic world. The spinors u and $\dot{u}$ are for the spin-up states and the spinors v and $\dot{v}$ are for spin-down states.

	Undotted	Dotted
Spin-up	u	$\dot{u}$
Spin-down	v	$\dot{v}$

equations (2.21) and (2.23). These groups of matrices, which form a representation of $SL(2, c)$, are not always unitary. This is because the generators K_i are not Hermitian. Thus the Hermitian conjugate of G is not necessarily the inverse. This two-by-two representation has a rich history and has been discussed extensively in the literature [1, 4, 5, 8, 9, 27]. The six independent real parameters of G make it possible to construct four-by-four matrices for Lorentz transformations applicable to the four-dimensional Minkowskian space [5, 21].

As we did in equation (2.29) for this four-dimensional Minkowskian space, we can form a two-by-two representation for the space–time four-vector as

$$[X] = \begin{pmatrix} t + z & x - iy \\ x + iy & t - z \end{pmatrix}, \tag{4.6}$$

where the determinant of this matrix is $t^2 - z^2 - x^2 - z^2$. Equation (4.6) remains invariant under the Hermitian transformation:

$$[X]' = G \, [X] \, G^\dagger. \tag{4.7}$$

This is a Lorentz transformation which can be explicitly written as

$$\begin{pmatrix} t' + z' & x' - iy' \\ x' + iy' & t' - z' \end{pmatrix} = \begin{pmatrix} \alpha & \beta \\ \gamma & \delta \end{pmatrix} \begin{pmatrix} t + z & x - iy \\ x + iy & t - z \end{pmatrix} \begin{pmatrix} \alpha^* & \gamma^* \\ \beta^* & \delta^* \end{pmatrix}. \tag{4.8}$$

Likewise, the two-by-two matrix for the four-momentum takes the form

$$[P] = \begin{pmatrix} p_0 + p_z & p_x - ip_y \\ p_x + ip_y & p_0 - p_z \end{pmatrix}, \tag{4.9}$$

with $p_0 = \sqrt{m^2 + p_z^2 + p_x^2 + p_2^2}$. The transformation property of equation (4.8) is applicable also to this energy–momentum four-vector.

In 1939 [33] Wigner considered the following three four-vectors

$$P_+ = \begin{pmatrix} 1 & 0 \\ 0 & 1 \end{pmatrix}, \qquad P_0 = \begin{pmatrix} 1 & 0 \\ 0 & 0 \end{pmatrix}, \qquad \text{and} \qquad P_- = \begin{pmatrix} 1 & 0 \\ 0 & -1 \end{pmatrix}. \tag{4.10}$$

The determinants of these matrices are 1, 0, and −1, respectively, corresponding to the four-momenta of massive, massless, and imaginary-mass particles. This is shown in table 4.2.

He then constructed the subgroups of the Lorentz group, i.e. Wigner's little groups, whose transformations leave these four-momenta invariant. Thus, the matrices of these little groups should satisfy

$$W \, P_i \, W^\dagger = P_i, \tag{4.11}$$

where $i = +, 0, -$. Since the momentum of the particle is fixed, these little groups define the internal space–time symmetries of the particle.

For all three cases, the momentum is invariant under rotations around the z-axis. Referring to table 2.1, this rotation matrix is

Table 4.2. The two-by-two matrix representation of Wigner momentum vectors together with the corresponding undotted and dotted transformation matrices. These four-momentum matrices have determinants that are positive, zero, and negative for massive, massless, and imaginary-mass particles, respectively.

Mass	Four-momentum	Transformation matrix	Dotted
Massive	$\begin{pmatrix} 1 & 0 \\ 0 & 1 \end{pmatrix}$	$\begin{pmatrix} \cos(\theta/2) & -\sin(\theta/2) \\ \sin(\theta/2) & \cos(\theta/2) \end{pmatrix}$	$\begin{pmatrix} \cos(\theta/2) & -\sin(\theta/2) \\ \sin(\theta/2) & \cos(\theta/2) \end{pmatrix}$
Massless	$\begin{pmatrix} 1 & 0 \\ 0 & 0 \end{pmatrix}$	$\begin{pmatrix} 1 & -\gamma \\ 0 & 1 \end{pmatrix}$	$\begin{pmatrix} 1 & 0 \\ \gamma & 1 \end{pmatrix}$
Imag. mass	$\begin{pmatrix} 1 & 0 \\ 0 & -1 \end{pmatrix}$	$\begin{pmatrix} \cosh(\lambda/2) & \sinh(\lambda/2) \\ \sinh(\lambda/2) & \cosh(\lambda/2) \end{pmatrix}$	$\begin{pmatrix} \cosh(\lambda/2) & -\sinh(\lambda/2) \\ -\sinh(\lambda/2) & \cosh(\lambda/2) \end{pmatrix}$

$$Z(\phi) = \begin{pmatrix} e^{i\phi/2} & 0 \\ 0 & e^{-i\phi/2} \end{pmatrix}. \tag{4.12}$$

Then

$$Z(\phi) \; P_i \; Z^\dagger(\phi) = P_i. \tag{4.13}$$

This means that the four-momentum remains invariant under rotations around the z-axis.

In addition, let us consider the transformation within the zx-plane. The matrix for these rotations is also given in table 2.1.

For the first case corresponding to a massive particle at rest, the requirement of the subgroup is

$$W \; P_+ \; W^\dagger = P_+. \tag{4.14}$$

This four-momentum remains invariant under rotations around the y-axis, whose transformation matrix is

$$R(\theta) = \begin{pmatrix} \cos(\theta/2) & -\sin(\theta/2) \\ \sin(\theta/2) & \cos(\theta/2) \end{pmatrix}. \tag{4.15}$$

This matrix together with $Z(\phi)$ also leads to rotation around the x-axis. Thus, Wigner's little group for the massive particle is the three-dimensional rotation subgroup of the Lorentz group generated by J_i given in equation (2.21).

For the second case of P_0, the triangular matrix of the form

$$T(\gamma) = \begin{pmatrix} 1 & -\gamma \\ 0 & 1 \end{pmatrix} \quad \text{or} \quad \dot{T}(\gamma) = \begin{pmatrix} 1 & 0 \\ \gamma & 1 \end{pmatrix} \tag{4.16}$$

satisfies the Wigner condition of equation (4.11). These matrices are not Hermitian, but rather

$$T^{-1}(\gamma) = \begin{pmatrix} 1 & \gamma \\ 0 & 1 \end{pmatrix} \quad \text{and} \quad T^\dagger(\gamma) = \begin{pmatrix} 1 & 0 \\ -\gamma & 1 \end{pmatrix} \tag{4.17}$$

and similarly for $\dot{T}(\gamma)$. In order to preserve the Lorentz properties of the boosted four-momentum, γ must be real. If we allow rotations around the z-axis, these triangular matrices become

$$T(\gamma e^{i\phi}) = \begin{pmatrix} 1 & -\gamma e^{i\phi} \\ 0 & 1 \end{pmatrix} \quad \text{or} \quad \dot{T}(\gamma e^{-i\phi}) = \begin{pmatrix} 1 & 0 \\ \gamma e^{-i\phi} & 1 \end{pmatrix}. \tag{4.18}$$

The T matrix is generated by

$$N_1 = K_1 - J_2 = \begin{pmatrix} 0 & i \\ 0 & 0 \end{pmatrix} \quad \text{and} \quad N_2 = K_2 + J_1 = \begin{pmatrix} 0 & 1 \\ 0 & 0 \end{pmatrix}. \tag{4.19}$$

The dotted matrix is generated by

$$\dot{N}_1 = \dot{K}_1 - J_2 = \begin{pmatrix} 0 & 0 \\ -i & 0 \end{pmatrix} \quad \text{and} \quad \dot{N}_2 = \dot{K}_2 + J_1 = \begin{pmatrix} 0 & 0 \\ 1 & 0 \end{pmatrix}. \tag{4.20}$$

Thus, the little group is generated by J_3, N_1, and N_2, or by J_3, $\dot{N}_1$, and $\dot{N}_2$. Hence, there are two sets of boost generators involved. These generators satisfy the following sets of commutation relations:

$$[N_1, N_2] = 0, \quad [N_1, J_3] = iN_2, \quad \text{and} \quad [N_2, J_3] = -iN_1, \tag{4.21}$$

and

$$[\dot{N}_1, \dot{N}_2] = 0, \quad [\dot{N}_1, J_3] = i\dot{N}_2, \quad \text{and} \quad [\dot{N}_2, J_3] = -i\dot{N}_1. \tag{4.22}$$

As usual J_3 is the generator of rotations and N_i generate translation-like transformations, where

$$D(\alpha, \beta) = D(\alpha, 0)D(0, \beta) = D(0, \beta)D(\alpha, 0). \tag{4.23}$$

Wigner in 1939 [33] observed that the first set given in equation (4.21) is the same as that of the generators for the two-dimensional Euclidean group with one rotation and two translations. The physical interpretation of the rotation is easy to understand. It is the helicity of the massless particle. On the other hand, the physics of the N_1 and N_2 matrices was not obvious, and the issue was not completely settled until 1990 [24]. They generate gauge transformations [13, 25, 31].

For the third case of P_-, the matrix of the form

$$S(\lambda) = \begin{pmatrix} \cosh(\lambda/2) & \sinh(\lambda/2) \\ \sinh(\lambda/2) & \cosh(\lambda/2) \end{pmatrix} \tag{4.24}$$

satisfies the Wigner condition of equation (4.11). This corresponds to the Lorentz boost along the x-direction generated by K_1 as shown in table 2.1. Because of the rotational symmetry around the z-axis, the Wigner condition is satisfied also by the boost along the y-axis.

The little group is thus generated by J_3, K_1, and K_2. These three generators,

$$[J_3, K_1] = iK_2, \qquad [J_3, K_2] = -iK_1, \qquad \text{and} \qquad [K_1, K_2] = -iJ_3, \qquad (4.25)$$

form the two-by-two representation of the little group $O(2, 1)$, which is the Lorentz group applicable to two space dimensions and one time dimension as discussed in section 3.1.3. The dotted matrices should satisfy the same set of commutation relations.

As far as rotations are concerned, the representation constructed from the Lie algebra of equations (2.21) and (2.23) is transformed in the same way as that of equation (4.5). However, the Lorentz boosts for the dotted representation are performed in opposite directions. This is shown in table 4.3.

There are interesting three-parameter subgroups of the Lorentz group. In 1939 [33] Wigner considered only those subgroups whose transformations leave the four-momentum of a given particle invariant. First of all, consider a massive particle at rest. The momentum of this particle is invariant under rotations in three-dimensional space. What happens for the massless particle that cannot be brought to a rest frame? We shall study this problem for understanding the space–time symmetry of neutrinos and photons.

4.2 Massive and massless particles

Indeed, the massive particle at rest remains invariant under rotations. Let us Lorentz-boost this particle along the z-direction. The boost matrix is given in table 2.1 and it takes the form

Table 4.3. Two-by-two representations of the Lorentz group. Rotations take the same form for both dotted and undotted representations, but boosts are performed in opposite directions.

Generators	Transformation matrix		Dotted
$J_1 = \frac{1}{2}\begin{pmatrix} 0 & 1 \\ 1 & 0 \end{pmatrix}$	$\begin{pmatrix} \cos(\theta/2) & i\sin(\theta/2) \\ i\sin(\theta/2) & \cos(\theta/2) \end{pmatrix}$	same	$\begin{pmatrix} \cos(\theta/2) & i\sin(\theta/2) \\ i\sin(\theta/2) & \cos(\theta/2) \end{pmatrix}$
$K_1 = \frac{1}{2}\begin{pmatrix} 0 & i \\ i & 0 \end{pmatrix}$	$\begin{pmatrix} \cosh(\lambda/2) & \sinh(\lambda/2) \\ \sinh(\lambda/2) & \cosh(\lambda/2) \end{pmatrix}$	inverse	$\begin{pmatrix} \cosh(\lambda/2) & -\sinh(\lambda/2) \\ -\sinh(\lambda/2) & \cosh(\lambda/2) \end{pmatrix}$
$J_2 = \frac{1}{2}\begin{pmatrix} 0 & -i \\ i & 0 \end{pmatrix}$	$\begin{pmatrix} \cos(\theta/2) & -\sin(\theta/2) \\ \sin(\theta/2) & \cos(\theta/2) \end{pmatrix}$	same	$\begin{pmatrix} \cos(\theta/2) & -\sin(\theta/2) \\ \sin(\theta/2) & \cos(\theta/2 \end{pmatrix}$
$K_2 = \frac{1}{2}\begin{pmatrix} 0 & 1 \\ -1 & 0 \end{pmatrix}$	$\begin{pmatrix} \cosh(\lambda/2) & -i\sinh(\lambda/2) \\ i\sinh(\lambda/2) & \cosh(\lambda/2) \end{pmatrix}$	inverse	$\begin{pmatrix} \cosh(\lambda/2) & i\sinh(\lambda/2) \\ -i\sinh(\lambda/2) & \cosh(\lambda/2) \end{pmatrix}$
$J_3 = \frac{1}{2}\begin{pmatrix} 1 & 0 \\ 0 & -1 \end{pmatrix}$	$\begin{pmatrix} e^{i\phi/2} & 0 \\ 0 & e^{-i\phi/2} \end{pmatrix}$	same	$\begin{pmatrix} e^{i\phi/2} & 0 \\ 0 & e^{-i\phi/2} \end{pmatrix}$
$K_3 = \frac{1}{2}\begin{pmatrix} i & 0 \\ 0 & -i \end{pmatrix}$	$\begin{pmatrix} e^{\eta/2} & 0 \\ 0 & e^{-\eta/2} \end{pmatrix}$	inverse	$\begin{pmatrix} e^{-\eta/2} & 0 \\ 0 & e^{\eta/2} \end{pmatrix}$

$$B(\eta) = \begin{pmatrix} e^{\eta/2} & 0 \\ 0 & e^{-\eta/2} \end{pmatrix}. \tag{4.26}$$

Then the momentum of the massive particle becomes

$$p_z = m \, \sinh(\eta), \tag{4.27}$$

where the boost parameter η is now related to the physical variables as

$$e^\eta = \frac{p_z + \sqrt{p_z^2 + m^2}}{m}. \tag{4.28}$$

This momentum remains invariant under rotations around the z-axis. The rotation matrix $Z(\phi)$ given in equation (4.12) commutes with the boost matrix $B(\eta)$ of equation (4.26).

The story is different for rotations around an axis perpendicular to the z-axis. Let us pick the rotation around the y-axis given in equation (4.15). This matrix becomes boosted to

$$B(\eta)R(\theta)B^{-1}(\eta) = \begin{pmatrix} \cos(\theta/2) & -e^\eta \sin(\theta/2) \\ e^{-\eta} \sin(\theta/2) & \cos(\theta/2) \end{pmatrix}. \tag{4.29}$$

According to equation (4.27), η becomes infinite as the mass becomes smaller. If we decide to keep all the quantities in equation (4.29) finite, the upper-right element $e^\eta \sin(\theta/2)$ must be finite. Let that be γ. The lower-left element then becomes $e^{-2\eta}\gamma$ which vanishes as η becomes infinite. As the angle θ becomes zero, the boosted rotation matrix becomes the triangular matrix

$$T = \begin{pmatrix} 1 & -\gamma \\ 0 & 1 \end{pmatrix} \quad \text{and} \quad \dot{T} = \begin{pmatrix} 1 & 0 \\ \gamma & 1 \end{pmatrix}, \tag{4.30}$$

which are the triangular Wigner matrices given in equation (4.16). When they are applied to the spinors given in table 4.1, u and $\dot{v}$ remain invariant, but $\dot{u}$ and v become changed as shown in table 4.4.

Here again, there is the rotational degree of freedom around the z-axis. The matrix of equation (4.29) is generalized into

$$\begin{pmatrix} e^{i\phi/2} & 0 \\ 0 & e^{-i\phi/2} \end{pmatrix} \begin{pmatrix} \cos(\theta/2) & -e^\eta \sin(\theta/2) \\ e^{-\eta} \sin(\theta/2) & \cos(\theta/2) \end{pmatrix} \begin{pmatrix} e^{-i\phi/2} & 0 \\ 0 & e^{i\phi/2} \end{pmatrix}, \tag{4.31}$$

which becomes

$$\begin{pmatrix} \cos(\theta/2) & -e^{i\phi}e^\eta \sin(\theta/2) \\ e^{-i\phi}e^{-\eta} \sin(\theta/2) & \cos(\theta/2) \end{pmatrix}. \tag{4.32}$$

In the large-η limit, this expression leads to the triangular matrices of equation (4.18).

Table 4.4. $T(\gamma)$ and $\dot{T}(\gamma)$ transformations on the spinors. Due to the parity invariance of the Lie algebra of the Lorentz group, we should consider the triangular matrices applicable to both u and v, and also to $\dot{u}$ and $\dot{v}$.

	$T(\gamma)$ with $+\eta$	$\dot{T}(\gamma)$ with $-\eta$
Spinors	$T(\gamma)u = u$	$\dot{T}(\gamma)u = u + \gamma v$
	$T(\gamma)v = v - \gamma u$	$\dot{T}(\gamma)v = v$
Dotted spinors	$T(\gamma)\dot{u} = \dot{u}$	$\dot{T}(\gamma)\dot{u} = \dot{u} + \gamma\dot{v}$
	$T(\gamma)\dot{v} = \dot{v} - \gamma\dot{u}$	$\dot{T}(\gamma)\dot{v} = \dot{v}$

We have now established that the triangular matrices T and $\dot{T}$ generate gauge transformations when applied to four-vectors. Let us go back to table 4.4. It is now possible to show that they perform gauge transformations on massless spin-half particles [14, 16], as well as on photons.

4.3 Polarization of massless neutrinos

In terms of the N_i and $\dot{N}_i$ the transformation matrices defined in equation (4.23) can be written as [16, 20]

$$D(\alpha, \beta) = e^{-i[\alpha N_1 + \beta N_2]} = \begin{pmatrix} 1 & \alpha - i\beta \\ 0 & 1 \end{pmatrix},$$

$$\dot{D}(\alpha, \beta) = e^{-i[\alpha \dot{N}_1 + \beta \dot{N}_2]} = \begin{pmatrix} 1 & 0 \\ -\alpha - i\beta & 1 \end{pmatrix}. \tag{4.33}$$

Since there are two set of spinors in $SL(2, c)$, we see that they are gauge-invariant in the sense that

$$D(\alpha, \beta)u = u \quad \text{and} \quad \dot{D}(\alpha, \beta)\dot{v} = \dot{v}. \tag{4.34}$$

However, if we carry out the explicit multiplication, these spinors are gauge-dependent in the sense that

$$D(\alpha, \beta)v = v + (\alpha - i\beta)u \quad \text{and} \quad \dot{D}(\alpha, \beta)\dot{u} = \dot{u} - (\alpha + i\beta)\dot{v}. \tag{4.35}$$

The gauge-invariant spinors of equation (4.34) appear as polarized neutrinos [14, 19, 23].

However, if we insist on gauge invariance of the world, massless spin-half particles are polarized, as shown in figure 4.1. The dotted spinor becomes left-handed, while the undotted spinor becomes right-handed [14, 16, 17]. Indeed, this is what we observe in the real world. Massless neutrinos and anti-neutrinos are left- and right-handed, respectively.

Yes, neutrinos have non-zero masses [20, 26], but they are so small compared to their momenta that they can be regarded as small corrections to their massless states. In other words, their massless states still play important roles in physics.

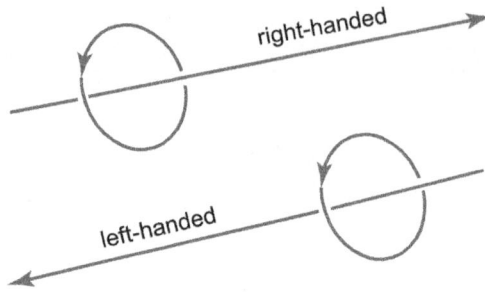

Figure 4.1. Polarization of massless neutrinos. This polarization is a consequence of gauge invariance [22].

4.3.1 Dirac spinors and massless particles

The Dirac equation is also applicable to massive particles. Here we will consider the massless particle as the limiting case of the massive particle by considering the large-momentum/zero-mass limit of the Dirac spinors.

Starting with the spin operators defined in equations (2.21) and (2.23), a boost applied on J_i along the z-direction will yield

$$J_i' = B(\eta) \, J_i \, B^{-1}(\eta). \tag{4.36}$$

This is a similarity transformation. Here the boost matrix is given by equation (4.26), where now the boost parameter becomes

$$e^\eta = \left(\frac{E+P}{E-P}\right)^{1/2}. \tag{4.37}$$

In the large-momentum or small-mass limit for a massive particle we obtain

$$e^\eta \to \frac{2E}{m}. \tag{4.38}$$

Using the similarity transformation of equation (4.36), J_3 is invariant, but J_1 and J_2 take the form

$$J_1' = \frac{1}{2}\begin{pmatrix} 0 & e^\eta \\ e^{-\eta} & 0 \end{pmatrix} \quad \text{and} \quad J_2' = \frac{i}{2}\begin{pmatrix} 0 & -e^\eta \\ e^{-\eta} & 0 \end{pmatrix}. \tag{4.39}$$

In the large-momentum or small-mass limit for a massive particle, we can obtain the N_i matrices of equation (4.20) as

$$N_1 = -\frac{m}{E}J_2' \quad \text{and} \quad N_2 = \frac{E}{m}J_1'. \tag{4.40}$$

Remembering that we have to consider both signs of the boost generators, the generators of $SL(2, c)$ can take the form

$$J_i = \frac{1}{2}\begin{pmatrix} \sigma_i & 0 \\ 0 & \sigma_i \end{pmatrix} \quad \text{and} \quad K_i = \frac{i}{2}\begin{pmatrix} \sigma_i & 0 \\ 0 & \sigma_i \end{pmatrix}, \tag{4.41}$$

which is applicable to Dirac wave functions in the Weyl representation [2, 21]. Using the gauge transformation matrices from equation (4.23), we can write

$$D(\alpha, \beta) = \begin{pmatrix} D(\alpha, \beta) & 0 \\ 0 & \dot{D}(\alpha, \beta) \end{pmatrix}. \tag{4.42}$$

This matrix is applicable to the Dirac spinors. To evaluate the result of applying the D matrix from equation (4.42), we first look at the eigenspinors given in equation (4.2) applied to a massive Dirac particle that is at rest. Thus we have

$$u(0) = \begin{pmatrix} u \\ \pm \dot{u} \end{pmatrix} \quad \text{and} \quad v(0) = \begin{pmatrix} \pm v \\ \dot{v} \end{pmatrix}, \tag{4.43}$$

where the positive and negative energy states are denoted by the $+$ and $-$ signs, respectively. If these spinors are boosted along the z-axis using the operator generated by K_3, then

$$u(\eta) = \begin{pmatrix} e^{+\eta/2}u \\ \pm e^{-\eta/2}\dot{u} \end{pmatrix} \quad \text{and} \quad v(\eta) = \begin{pmatrix} \pm e^{-\eta/2}v \\ e^{+\eta/2}\dot{v} \end{pmatrix}. \tag{4.44}$$

In the large-momentum/zero-mass limit, the large components, $e^{(+\eta/2)}$, are, according to equation (4.34), gauge-invariant, while the small components, according to equation (4.35), are gauge-dependent. This again shows that non-zero-mass, spin-$\frac{1}{2}$ particles are not invariant under gauge transformations. Furthermore, in this limit, the spinors of equation (4.44) can be renormalized as

$$u = \begin{pmatrix} u \\ 0 \end{pmatrix} \quad \text{and} \quad v = \begin{pmatrix} 0 \\ \dot{v} \end{pmatrix}. \tag{4.45}$$

It is clear that the D transformation leaves these spinors invariant. It is this invariance, as shown before, that is responsible for the polarization of neutrinos [14, 19].

Additionally, one could interpret the results of equation (4.44) in terms of $E(2)$ translations on free Weyl neutrino states. In this case, the gauge-invariant transformations leave the left-handed neutrino invariant, but translate the right-handed neutrino into a linear combination of left-handed and right-handed neutrinos [2, 14]. These coupled states could have implications requiring that in a constant electric and magnetic field neutrinos should acquire a small effective mass [2].

4.4 Scalars, vectors, tensors, and the polarization of photons

We are quite familiar with the process of constructing one spin-0 state and three spin-1 states from two spinors. Since each spinor has two states, there are four states if combined.

In the Lorentz-covariant world, for each spin-1/2 particle, there are two additional two-component spinors coming from the dotted representation [5, 6, 11, 21, 28].

There are thus four states. If two spinors are combined, there are 16 states. They can be partitioned into the following states:

1. Scalar with one state.
2. Pseudo-scalar with one state.
3. Four-vector with four states.
4. Axial vector with four states.
5. Second-rank tensor with six states.

If the particle is at rest and we consider only the undotted spinors, we can explicitly construct the combinations

$$\frac{1}{\sqrt{2}}(uv - vu), \tag{4.46}$$

for the spin-zero state and

$$uu, \qquad \frac{1}{\sqrt{2}}(uv + vu), \qquad \text{and} \qquad vv, \tag{4.47}$$

to obtain the spin-1 states as we mentioned earlier. This results in four bilinear states. In the $SL(2, c)$ regime, we must include both dotted and undotted spinors. Then there are sixteen independent bilinear combinations, as given in table 4.5.

Among the bilinear combinations given in table 4.5, the following two equations are invariant under rotations and also under boosts:

$$S = \frac{1}{\sqrt{2}}(uv - vu) \quad \text{and} \quad \dot{S} = -\frac{1}{\sqrt{2}}(\dot{u}\dot{v} - \dot{v}\dot{u}). \tag{4.48}$$

These are thus scalars in the Lorentz-covariant world. Are they the same or different? Let us consider the following combinations

$$S_+ = \frac{1}{\sqrt{2}}(S + \dot{S}) \quad \text{and} \quad S_- = \frac{1}{\sqrt{2}}(S - \dot{S}). \tag{4.49}$$

Under the dot conjugation, S_+ remains invariant, but S_- changes sign. From this we see that the dot conjugation corresponds to space inversion or the parity operation. Thus, S_+ is a scalar, while S_- is called a pseudo-scalar.

Table 4.5. Sixteen combinations of the $SL(2, c)$ spinors. In the $SU(2)$ regime, there are two spinors leading to four bilinear forms. In the $SL(2, c)$ world, there are two undotted and two dotted spinors. These four-spinors lead to sixteen independent bilinear combinations [5].

Spin-1			Spin-0
$uu,$	$\frac{1}{\sqrt{2}}(uv + vu),$	$vv,$	$\frac{1}{\sqrt{2}}(uv - vu)$
$\dot{u}\dot{u},$	$\frac{1}{\sqrt{2}}(\dot{u}\dot{v} + \dot{v}\dot{u}),$	$\dot{v}\dot{v},$	$\frac{1}{\sqrt{2}}(\dot{u}\dot{v} - \dot{v}\dot{u})$
$u\dot{u},$	$\frac{1}{\sqrt{2}}(u\dot{v} + v\dot{u}),$	$v\dot{v},$	$\frac{1}{\sqrt{2}}(u\dot{v} - v\dot{u})$
$\dot{u}u,$	$\frac{1}{\sqrt{2}}(\dot{u}v + \dot{v}u),$	$\dot{v}v,$	$\frac{1}{\sqrt{2}}(\dot{u}v - \dot{v}u)$

4.4.1 Four-vectors

Let us rewrite the expression for the space–time four-vector given in equation (4.6) as

$$\begin{pmatrix} t + z & x - iy \\ x + iy & t - z \end{pmatrix},$$ (4.50)

which, under the parity operation, becomes

$$\begin{pmatrix} t - z & - x + iy \\ - x - iy & t + z \end{pmatrix}.$$ (4.51)

The off-diagonal elements undergo sign changes, and the diagonal elements become interchanged. We can now construct the four-vectors such as equation (4.50) and its *dot* conjugation as

$$V \simeq \begin{pmatrix} \dot{v}u - u\dot{v} & \dot{v}v - v\dot{v} \\ u\dot{u} - \dot{u}u & v\dot{u} - \dot{u}v \end{pmatrix} \quad \text{and} \quad \dot{V} \simeq \begin{pmatrix} v\dot{u} - \dot{u}v & v\dot{v} - \dot{v}v \\ \dot{u}u - u\dot{u} & \dot{v}u - u\dot{v} \end{pmatrix},$$ (4.52)

respectively.

Similarly, we write the electromagnetic four-vector as

$$A = \begin{pmatrix} A_0 + A_z & A_x - iA_y \\ A_x + iA_y & A_0 - A_z \end{pmatrix}.$$ (4.53)

If boosted along the z-direction, this matrix becomes

$$A = \begin{pmatrix} (A_0 + A_z)e^{\eta} & A_x - iA_y \\ A_x + iA_y & (A_0 - A_z)e^{-\eta} \end{pmatrix}.$$ (4.54)

We can then make the Wigner excursion as illustrated in figure 3.4, which transforms this matrix for a massive particle at rest to that of a massless particle with the same energy. The net result is

$$A = \begin{pmatrix} A_0 + A_z & A_x - iA_y \\ A_x + iA_y & 0 \end{pmatrix},$$ (4.55)

resulting in $A_0 = A_z$, which is widely known as the Lorentz condition.

If we perform the $T(\gamma)$ transformation on u and v, while we use $\dot{T}(\gamma)$ on $\dot{u}$ and $\dot{v}$, the A matrix becomes

$$A + 2\gamma \begin{pmatrix} 0 & A_0 \\ A_0 & 0 \end{pmatrix},$$ (4.56)

This results in the addition of $2\gamma A_0$ to A_x. It is a translation in the plane of A_x and A_y.

On the other hand, if we perform the $\dot{T}(\gamma)$ transformation u and v, while we use $T(\gamma)$ on $\dot{u}$ and $\dot{v}$, A becomes

$$A - 2\gamma \begin{pmatrix} A_x & 0 \\ 0 & 0 \end{pmatrix}. \tag{4.57}$$

The transformations for those triangular matrices are given in table 4.4.

The question next is which to choose between equations (4.56) and (4.57) for our transformation. In order to decide, let us go to the transformation rule given in equation (4.8) for the four-vector, and apply the triangular matrix $T(\gamma)$ as the Lorentz transformation matrix, resulting in the matrix multiplication

$$\begin{pmatrix} 1 & -\gamma \\ 0 & 1 \end{pmatrix} \begin{pmatrix} A_0 + A_z & A_x - iA_y \\ A_x + iA_y & 0 \end{pmatrix} \begin{pmatrix} 1 & 0 \\ -\gamma & 1 \end{pmatrix}, \tag{4.58}$$

which becomes

$$A - 2\gamma \begin{pmatrix} A_x & 0 \\ 0 & 0 \end{pmatrix}. \tag{4.59}$$

This form is the same as the form given in equation (4.57). Thus, we choose the transformation of equation (4.57) as the $T(\gamma)$ and $\dot{T}(\gamma)$ transformations applicable to the spinors.

In both equations (4.57) and (4.59), A_x and A_y remain invariant while there is an addition to $A_0 = A_z$. The $T(\gamma)$ matrices therefore lead to a gauge transformation.

What we have done so far can be rotated around the z-axis. Then, γ is replaced by $\gamma e^{i\phi}$. The transformed A of equation (4.57) becomes

$$A - 2\gamma \begin{pmatrix} A_x \cos\phi + A_y \sin\phi & 0 \\ 0 & 0 \end{pmatrix}. \tag{4.60}$$

It is possible to reach the same conclusion using the four-by-four formulation of the Lorentz group. This larger representation contains geometries leading to equations (4.56) and (4.60) [13, 24, 25, 31]. We now know from equation (4.60) that $T(\gamma e^{i\phi})$ performs a gauge transformation.

Let us go back to the limiting process discussed in section 4.2. According to section 4.2, the transverse rotational degrees of freedom collapse into one gauge degree of freedom in the infinite-momentum or zero-mass limit. This aspect was observed first by Han *et al* in 1983 [15], and its geometry was given by Kim and Wigner in 1990 [24]. The most recent version of this geometry was given by the present authors in 2017 [4].

The matrices given in this section were given in our earlier paper on this subject [4]. However, as we go deeper into the problem by starting from the transformation property of each spinor, it is inevitable to make a number of minus-sign adjustments. These changes do not alter the conclusions given there or those given here.

4.4.2 Second-rank tensor

There are also bilinear spinors, which are both dotted or both undotted. We are interested in two sets of three quantities satisfying the $O(3)$ symmetry. They should therefore transform like

$$(x + iy)/\sqrt{2}, \qquad (x - iy)/\sqrt{2}, \qquad \text{and} \qquad z, \qquad (4.61)$$

which are like

$$uu, \qquad vv, \qquad \text{and} \qquad (uv + vu)/\sqrt{2}, \qquad (4.62)$$

respectively, in the $O(3)$ regime. Since the dot conjugation is the parity operation, they are like

$$-\dot{u}\dot{u}, -\dot{v}\dot{v} \quad \text{and} \quad - (\dot{u}\dot{v} + \dot{v}\dot{u})/\sqrt{2}. \qquad (4.63)$$

In other words,

$$(uu\dot{)} = -\dot{u}\dot{u} \quad \text{and} \quad (vv\dot{)} = -\dot{v}\dot{v}. \qquad (4.64)$$

We noticed a similar sign change in equation (4.51).

In order to construct the z-component in this $O(3)$ space, let us first consider

$$f_z = \frac{1}{2}[(uv + vu) - (\dot{u}\dot{v} + \dot{v}\dot{u})] \quad \text{and} \quad g_z = \frac{1}{2i}[(uv + vu) + (\dot{u}\dot{v} + \dot{v}\dot{u})]. \quad (4.65)$$

Here, f_z and g_z are, respectively, symmetric and anti-symmetric under the dot conjugation or the parity operation. These quantities are invariant under the boost along the z-direction. They are also invariant under rotations around this axis, but they are not invariant under boosts along or rotations around the x- or y-axis. They are different from the scalars given in equation (4.48).

Next, in order to construct the x- and y-components, we start with $f_\pm$ and $g_\pm$ as

$$f_+ = \frac{1}{\sqrt{2}}(uu - \dot{u}\dot{u}) \qquad \text{and} \qquad f_- = \frac{1}{\sqrt{2}}(vv - \dot{v}\dot{v}),$$
$$g_+ = \frac{1}{\sqrt{2}\,i}(uu + \dot{u}\dot{u}) \qquad \text{and} \qquad g_- = \frac{1}{\sqrt{2}\,i}(vv + \dot{v}\dot{v}). \qquad (4.66)$$

Then

$$f_x = \frac{1}{\sqrt{2}}(f_+ + f_-) = \frac{1}{2}[(uu + vv) - (\dot{u}\dot{u} + \dot{v}\dot{v})],$$
$$f_y = \frac{1}{\sqrt{2}\,i}(f_+ - f_-) = \frac{1}{2i}[(uu - vv) - (\dot{u}\dot{u} - \dot{v}\dot{v})], \qquad (4.67)$$

and

$$g_x = \frac{1}{\sqrt{2}}(g_+ + g_-) = \frac{1}{2i}[(uu + vv) + (\dot{u}\dot{u} + \dot{v}\dot{v})],$$
$$g_y = \frac{1}{\sqrt{2}\,i}(g_+ - g_-) = -\frac{1}{2}[(uu - vv) + (\dot{u}\dot{u} - \dot{v}\dot{v})]. \qquad (4.68)$$

Here, f_x and f_y are anti-symmetric under dot conjugation, while g_x and g_y are symmetric.

Furthermore, f_z, f_x, and f_y of equations (4.65) and (4.67) transform like a three-dimensional vector. The same can be said for g_i of equations (4.65) and (4.68). Thus, they can be grouped to form a second-rank tensor,

$$\begin{pmatrix} 0 & -f_z & -f_x & -f_y \\ f_z & 0 & -g_y & g_x \\ f_x & g_y & 0 & -g_z \\ f_y & -g_x & g_z & 0 \end{pmatrix}, \tag{4.69}$$

whose Lorentz transformation properties are well known. The g_i components change their signs under space inversion, while the f_i components remain invariant. They are like the electric and magnetic fields, respectively.

If the system is Lorentz-boosted, f_i and g_i can be computed from table 4.5. We are now interested in the symmetry of photons by taking the Wigner excursion as illustrated in figure 3.4.

Thus, we keep only the terms that become larger for larger values of η. Thus,

$$f_x \to \frac{1}{2}(uu - v\dot{v}) \quad \text{and} \quad f_y \to \frac{1}{2i}(uu + v\dot{v}),$$

$$g_x \to \frac{1}{2i}(uu + v\dot{v}) \quad \text{and} \quad g_y \to -\frac{1}{2}(uu - v\dot{v}), \tag{4.70}$$

in the massless limit.

Then, the tensor of equation (4.69) becomes [3]

$$\begin{pmatrix} 0 & 0 & -E_x & -E_y \\ 0 & 0 & -B_y & B_x \\ E_x & B_y & 0 & 0 \\ E_y & -B_x & 0 & 0 \end{pmatrix}, \tag{4.71}$$

with

$$E_x \simeq \frac{1}{2}(uu - v\dot{v}) \quad \text{and} \quad E_y \simeq \frac{1}{2i}(uu + v\dot{v}),$$

$$B_x \simeq \frac{1}{2i}(uu + v\dot{v}) \quad \text{and} \quad B_y \simeq -\frac{1}{2}(uu - v\dot{v}). \tag{4.72}$$

The four-by-four matrix of equation (4.71) is consistent with the Maxwell tensor given in equation (4.1). The electric and magnetic field components are perpendicular to each other. Furthermore,

$$B_x = E_y \quad \text{and} \quad B_y = -E_x. \tag{4.73}$$

In order to address the symmetry of photons, let us go back to equation (4.66). After the Wigner excursion,

$$B_+ \simeq E_+ \simeq uu \quad \text{and} \quad B_- \simeq E_- \simeq \dot{v}\dot{v}. \tag{4.74}$$

The gauge transformations applicable to u and $\dot{v}$ are the two-by-two matrices given in equation (4.16). Both u and $\dot{v}$ are invariant under gauge transformations, while $\dot{u}$ and v are not as shown in table 4.4. The B_+ and E_+ are for the photon spin along the z-direction, while B_- and E_- are for the opposite direction. Although equation (4.74) is only for the field propagating along the z-direction, it can be rotated to suit an arbitrary direction [3].

It is important to note that the second-rank tensor in equation (4.69) is quite different from that given in equation (4.71). The former is for massive particles and forms a representation space for the $O(3)$-like little group. For massless particles the only gauge-invariant products are uu and $\dot{v}\dot{v}$. Therefore, only these products are accommodated as the components of the Maxwell tensor given in equation (4.71).

Most of the formulae presented in this subsection were given in our previous publication [4]. In the present subsection, we have given a more precise definition of the zero-mass limit in terms of the Wigner excursion as illustrated in figure 3.4.

4.4.3 Higher spins

Since Wigner's original book of 1931 [32, 34], the rotation group, without Lorentz transformations, has been extensively discussed in the literature [7, 11, 12]. One of the main issues was how to construct the most general spin state from the two-component spinors for the spin-1/2 particle.

Since there are two states for the spin-1/2 particle, four states can be constructed from two spinors, leading to one state for the spin-0 state and three spin-1 states. With three spinors, it is possible to construct four spin-3/2 states and two spin-1/2 states, resulting in six states. This partition process is much more complicated [10, 18] for the case of three spinors. Yet, this partition process is possible for all higher spin states.

In the Lorentz-covariant world, there are four states for each spin-1/2 particle. With two spinors, we end up with sixteen (4×4) states, and they are tabulated in table 4.5. There should be 64 states for three spinors and 256 states for four-spinors. We now know how to Lorentz-boost those spinors. We also know that the transverse rotations become gauge transformations in the limit of zero-mass or infinite-η. It is thus possible to bundle all of them into the table given in figure 4.2.

In the relativistic regime, we are interested in photons and gravitons. As was noted in sections 4.4.1 and 4.4.2, the observable components are invariant under gauge transformations. They are also the terms that become largest for large values of η.

We have seen in section 4.4.2 that the photon state consists of uu and $\dot{v}\dot{v}$ for those whose spins are parallel and anti-parallel to the momentum, respectively. Thus, for spin-2 gravitons, the states must be $uuuu$ and $\dot{v}\dot{v}\dot{v}\dot{v}$, respectively.

In his effort to understand photons and gravitons, Weinberg constructed his states for massless particles [29], in particular photons and gravitons [30]. He started with the conditions

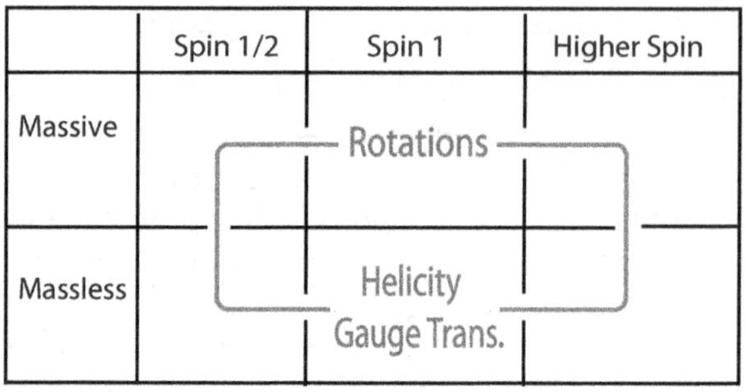

	Spin 1/2	Spin 1	Higher Spin
Massive		Rotations	
Massless		Helicity Gauge Trans.	

Figure 4.2. Unified picture of massive and massless particles. The gauge transformation is a Lorentz-boosted rotation matrix and is applicable to all massless particles. It is possible to construct higher spin states starting from the four states of the spin-1/2 particle in the Lorentz-covariant world [4].

$$N_1|\text{state}\rangle = 0 \quad \text{and} \quad N_2|\text{state}\rangle = 0, \tag{4.75}$$

where N_1 and N_2 are defined in equation (4.19). Since they are now known as the generators of gauge transformations, Weinberg's states are gauge-invariant states. Thus, uu and $\dot{v}\dot{v}$ are Weinberg's states for photons, and $uuuu$ are $\dot{v}\dot{v}\dot{v}\dot{v}$ are Weinberg's states for gravitons.

References

[1] Bargmann V 1947 Irreducible unitary representations of the Lorentz group *Ann. Math.* **48** 568
[2] Barut A O and McEwan J 1986 The four states of the Massless neutrino with Pauli coupling by spin-gauge invariance *Lett. Math. Phys.* **11** 67–72
[3] Başkal S and Kim Y S 1997 Little groups and Maxwell-type tensors for massive and massless particles *Europhys. Lett.* **40** 375–80
[4] Başkal S, Kim Y S, and Noz M E 2017 Loop representation of Wigner's little groups *Symmetry* **9** 97–118
[5] Başkal S, Kim Y S, and Noz M E 2015 *Physics of the Lorentz Group* (Bristol: IOP Publishing)
[6] Beresteckij V B, Lifshitz E M, Pitaevskij L P, and Landau L D 1982 *Quantum Electrodynamics Course of Theoretical Physics* vol 4 2nd edn (Amsterdam: Butterworth-Heinemann)
[7] Condon E U and Shortley G H 1951 *The Theory of Atomic Spectra* (Cambridge: Cambridge University Press)
[8] Dirac P A M 1945 Application of quaternions to Lorentz transformations *Proc. R. Irish Acad. Sect.* A **50** 261–70
[9] Dlugunovich V A and Kurochkin Y A 2009 Vector parameterization of the Lorentz group transformations and polar decomposition of Mueller matrices *Opt. Spectrosc.* **107** 294–8
[10] Feynman R P, Kislinger M, and Ravndal F 1971 Current matrix elements from a relativistic quark model *Phys. Rev.* D **3** 2706–32
[11] Gel'fand I M, Minlos R A, and Shapiro Z Y 2018 *Representations of the Rotation and Lorentz Groups and Their Applications* (Mineola, NY: Dover)
[12] Hamermesh M 1989 *Group Theory and Its Application to Physical Problems* (*Dover Books on Physics and Chemistry*) (New York: Dover)

[13] Han D and Kim Y S 1981 Little group for photons and gauge transformations *Am. J. Phys.* **49** 348–51

[14] Han D, Kim Y S, and Son D 1982 *E*(2)-like little group for massless particles and neutrino polarization as a consequence of gauge invariance *Phys. Rev.* D **26** 3717–25

[15] Han D, Kim Y S, and Son D 1983 Gauge transformations as Lorentz-boosted rotations *Phys. Lett.* B **131** 327–9

[16] Han D, Kim Y S, and Son D 1986 Eulerian parametrization of Wigner's little groups and gauge transformations in terms of rotations in two-component spinors *J. Math. Phys.* **27** 2228–35

[17] Han D, Kim Y S, and Son D 1986 Photons, neutrinos, and gauge transformations *Am. J. Phys.* **54** 818–21

[18] Hussar P E, Kim Y S, and Noz M E 1980 Three-particle symmetry classifications according to the method of Dirac *Am. J. Phys.* **48** 1038–42

[19] Kim Y S 2002 Neutrino polarization as a consequence of gauge invariance *Czech. J. Phys.* **52** C353–60

[20] Kim Y S, Maguire Jr G Q, and Noz M E 2016 Do small-mass neutrinos participate in gauge transformations? *Adv. High Energy Phys.* **2016** 1–7

[21] Kim Y S and Noz M E 1986 *Theory and Applications of the Poincaré Group* (Dordrecht: Springer)

[22] Kim Y S and Noz M E 2018 *New Perspectives on Einstein's E = mc²* (Singapore: World Scientific)

[23] Kim Y S and Wigner E P 1987 Cylindrical group and massless particles *J. Math. Phys.* **28** 1175–9

[24] Kim Y S and Wigner E P 1990 Space–time geometry of relativistic particles *J. Math. Phys.* **31** 55–60

[25] Kupersztych J 1976 Is there a link between gauge invariance, relativistic invariance and electron spin? *Nuovo Cim.* B **31** 1–11

[26] Mohapatra R N and Smirnov A Y 2006 Neutrino mass and new physics *Annu. Rev. Nucl. Part. Sci.* **56** 569–628

[27] Naimark M A 1954 Linear representation of the Lorentz group *Usp. Mat. Nauk* **9** 19–93
Naimark M A 1957 Linear representation of the Lorentz group *Am. Math. Soc. Transl. Ser.* **2** 379458 (Engl. transl.)
Naimark M A 1964 *Linear Representations of the Lorentz Group (International Series of Monographs in Pure and Applied Mathematics)* vol 63 (Oxford: Pergamon) (Engl. transl.)

[28] Weinberg S 1964 Feynman rules for any spin *Phys. Rev.* B **133** B1318–32

[29] Weinberg S 1964 Feynman rules for any spin. II. Massless particles *Phys. Rev.* B **134** B882–96

[30] Weinberg S 1964 Photons and gravitons in *S*-matrix theory: derivation of charge conservation and equality of gravitational and inertial mass *Phys. Rev.* B **135** B1049–56

[31] Weinberg S 1995 *The Quantum Theory of Fields: Foundations* vol 1 (Cambridge: Cambridge University Press)

[32] Wigner E P 1931 *Gruppentheorie und ihre Anwendung auf die Quantenmechanik der Atomspektren* (Braunschweig: Springer)

[33] Wigner E P 1939 On unitary representations of the inhomogeneous Lorentz group *Ann. Math.* **40** 149–204

[34] Wigner E P 1959 *Group Theory and its Application to the Quantum Mechanics of Atomic Spectra* (New York: Academic)

IOP Publishing

Mathematical Devices for Optical Sciences

Sibel Başkal, Young S Kim, and Marilyn E Noz

Chapter 5

Wigner functions

In chapter 1 we noted that there are several different representations of quantum mechanics. We showed that a convenient representation for non-pure mixed states was given by the density matrix formalism. Here we discuss another representation of quantum mechanics which is also convenient for mixed states, but regards both the position and momentum variables as c-numbers. This makes it possible to formulate quantum mechanics in phase space. This form will be referred to as the phase-space picture of quantum mechanics.

From chapter 1 we recall that the central role in the Schrödinger picture of quantum mechanics is played by the wave function. Hence, the phase-space distribution function introduced by Wigner [22] is the starting point for this phase-space picture of quantum mechanics. Widely known as the Wigner function, it is constructed from the Schrödinger wave function through the density matrix, a function of both position and momentum variables. Since it is not possible in quantum mechanics to determine the position and momentum variables simultaneously, the question is how can we represent the uncertainty principle in this phase-space picture of quantum mechanics? The position–momentum uncertainty is defined in terms of the area in phase space.

In section 5.1 we study the basic properties of the Wigner distribution function and give illustrative examples. We show that a very accurate description of the uncertainty relation for spreading wave packets is given by the phase-space picture of quantum mechanics.

In sections 5.2 and 5.3 the time dependence of the Wigner function is discussed. The wave packet spread is shown to preserve the amount of uncertainty.

In sections 5.4 and 5.5 we illustrate properties of the Wigner function in phase space by using the one-dimensional harmonic oscillator. The area-preserving transformations in phase space are discussed in detail. The area-preserving transformations in phase space are the transformations which preserve the amount of uncertainty.

doi:10.1088/2053-2563/aafe78ch5

In section 5.6 it is shown that it is possible to construct the density matrix from the Wigner function. Wigner functions and density matrices are needed in quantum mechanics when there are variables that are not measured. Those variables are treated statistically. In section 5.7 the density matrix allows us to calculate the entropy increase, while the Wigner function allows us to study the statistical uncertainty to be added to the quantum uncertainty.

5.1 Basic properties of the Wigner phase-space distribution function

For a wave function which depends on $x_1, x_2, \ldots, x_n$, and t, the density matrix is

$$\rho(x_1, \ldots, x_n; x_1', \ldots, x_n'; t) = \sum_k w_k \psi_k(x_1, \ldots, x_n, t)\psi_k^*(x_1', \ldots, x_n', t). \quad (5.1)$$

This is a straight-forward generalization of the single variable-density matrix discussed in chapter 1, with

$$\sum_k w_k = 1. \quad (5.2)$$

Thus the Wigner phase-space distribution function is defined as [12, 22]

$$W(x_1, \ldots, x_n; p_1, \ldots, p_n; t) = \left(\frac{1}{\pi}\right)^n \int e^{2i(p_1 y_1 + \cdots + p_n y_n)}$$
$$\times \rho(x_1 - y_1, \ldots, x_n - y_n; x_1 + y_1, \ldots, x_n + y_n; t)dy_1 dy_2 \cdots dy_n, \quad (5.3)$$

where x_i and p_i are c-numbers. Therefore, the Wigner function is defined over the $2n$-dimensional phase space. When the wave function is $\psi(x_1, \ldots, x_n, t)$ the system is in a pure state and

$$W(x_1, \ldots, x_n; p_1, \ldots, p_n; t) = \left(\frac{1}{\pi}\right)^n \int e^{2i(p_1 y_1 + \cdots + p_n y_n)}$$
$$\times \psi^*(x_1 + y_1, \ldots, x_n + y_n, t)\psi(x_1 - y_1, \ldots, x_n - y_n, t)dy_1 dy_2 \cdots dy_n. \quad (5.4)$$

When the system depends only on one pair of x and p variables, the pure-state Wigner function is simplest.

Starting with this simplest form Wigner function is the easiest way to study the properties of the Wigner function. In this simplest case, the Wigner function takes the form

$$W(x, p, t) = \frac{1}{\pi} \int e^{2ipy}\psi^*(x + y, t)\psi(x - y, t)dy. \quad (5.5)$$

If the wave function $\psi(x)$ is time-independent—the Wigner function does not depend on time. Indeed, the most frequently seen Wigner function in the literature takes the form

$$W(x, p) = \frac{1}{\pi} \int e^{2ipy}\psi^*(x + y)\psi(x - y)dy. \quad (5.6)$$

We first study this simplest form and then generalize its properties to the more complicated expressions given in equation (5.4).

1. When we integrate $W(x, p)$ over p, the quantum probability distribution in x is

$$\int W(x, p)dp = |\psi(x)|^2. \tag{5.7}$$

Similarly, if integrated over x, $W(x, p)$ gives the probability distribution in the momentum coordinate:

$$\int W(x, p)dx = |\chi(p)|^2, \tag{5.8}$$

where $\chi(p)$ is the momentum wave function derivable from $\psi(x)$ through

$$\chi(p) = \left(\frac{1}{2\pi}\right)^{1/2} \int e^{-ipx}\psi(x)dx. \tag{5.9}$$

Throughout equations (5.3)–(5.6), we note that p was introduced as a Fourier-transform parameter in the definition of the Wigner function. Based on the properties of equations (5.7) and (5.8), we can regard p as the momentum variable. In the Wigner phase space, both x and p are c-numbers. This aspect is different from the case in either the Schrödinger or Heisenberg picture of quantum mechanics. As we shall see in section 5.2, the uncertainty principle therefore has to be stated in a manner different from the conventional Schrödinger or Heisenberg picture.

2. The Wigner function can also be written as

$$W(x, p) = \left(\frac{1}{\pi}\right) \int e^{-2ixz}\chi^*(p + z)\chi(p - z)dz, \tag{5.10}$$

since we have

$$\psi(x) = \left(\frac{1}{2\pi}\right)^{1/2} \int e^{ipx}\chi(p)dp. \tag{5.11}$$

Starting from the momentum wave function, it is thus possible to define the Wigner function. It is also possible to reproduce the probability distributions of equations (5.7) and (5.8) from the above expression.

3. If an observation is made that changes the system's state vector $\psi(x)$ to $\phi(x)$, the probability of this result is $|(\psi, \phi)|^2$. This is the absolute square of the scalar product of the two state vectors. In terms of the Wigner functions, the transition probability can be written as

$$|(\psi, \phi)|^2 = 2\pi \int W_\psi(x, p)W_\phi(x, p)dxdp. \tag{5.12}$$

From this expression it is clear that the Wigner function cannot be positive everywhere in phase space. The above expression must vanish if ψ and ϕ are orthogonal. This is not possible if both of the Wigner functions on the right-hand side are positive everywhere in phase space.

4. The expectation values of two dynamical operators $A(x)$ and $B(p)$ depending only on x and p, respectively, are

$$(\psi(x),\, A(x)\psi(x)) = \int A(x)W(x,\, p)dxdp \qquad (5.13)$$

and

$$(\chi(p),\, B(p)\chi(p)) = \int B(p)W(x,\, p)dxdp. \qquad (5.14)$$

We know that in the Schrödinger picture $A(x)$ and $B(p)$ do not always commute with each other. Thus the expectation value of the product $A(x)B(p)$ does not take a simple form. This problem will be discussed in section 5.7.

5. As in the case of the distribution functions in classical phase space since both x and p are c-numbers, it is possible to perform canonical transformations in phase space. As was discussed in the paragraph following equation (5.12), the Wigner function can become negative in phase space. Therefore, the Wigner function, unlike the Liouville distribution function in classical mechanics, is not a probability distribution function.

It is possible to generalize the above properties (1)–(5) for a single pair of canonical variables to the Wigner function of many pairs of variables. The time-dependence of the Wigner function and the mixed-state Wigner function will be addressed in sections 5.2 and 5.5, respectively.

5.2 Time dependence of the Wigner function

Starting from the time-dependent Schrödinger equation it is possible to derive a differential equation for the Wigner function:

$$i\frac{\partial}{\partial t}\psi(x,\, t) = -\left(\frac{1}{2m}\right)\left(\frac{\partial}{\partial x}\right)^2\psi(x,\, t) + V(x)\psi(x,\, t), \qquad (5.15)$$

where m is the mass of the particle and $V(x)$ is the potential. This is because the Wigner function can be constructed from a time-dependent wave function. Using the definition of the Wigner function we have

$$\frac{\partial}{\partial t}W(x,\, p,\, t) = \left(\frac{i}{2\pi m}\right)\int e^{2ipy}\left\{\psi^*(x+y,\, t)\left[\left(\frac{\partial}{\partial x}\right)^2\psi(x-y,\, t)\right]\right.$$

$$\left. -\left[\left(\frac{\partial}{\partial x}\right)^2\psi^*(x+y,\, t)\right]\psi(x-y,\, t)\right\}dy$$

$$+\frac{i}{\pi}\int e^{2ipy}(V(x+y) - V(x-y)) \qquad (5.16)$$

$$\times \psi^*(x+y,\, t)\psi(x-y,\, t)dy.$$

Expanding the potential factor $(V(x+y) - V(x-y))$ results in

$$(V(x + y) - V(x - y)) = \sum_{n=0}^{\infty} \frac{2}{(2n + 1)!} \left\{ \left(\frac{\partial}{\partial x} \right)^{2n+1} V(x) \right\} y^{2n+1}. \tag{5.17}$$

Thus,

$$\frac{\partial}{\partial x} W(x, p, t) = -\left(\frac{i}{2\pi m} \right) \frac{\partial}{\partial x} \int e^{2ipy} \frac{\partial}{\partial y} (\psi^*(x + y, t)\psi(x - y, t)) dy$$

$$+ \left(\frac{i}{\pi} \right) \sum_{n=0}^{\infty} \frac{2}{(2n + 1)!} \left\{ \left(\frac{\partial}{\partial x} \right)^{2n+1} V(x) \right\} \tag{5.18}$$

$$\times \int e^{2ipy} y^{2n+1} \psi^*(x + y, t)\psi(x - y, t) dy.$$

In the above expression y^{2n+1} and $\partial/\partial y$ can be replaced by $(i/2)\partial/\partial p$ and $2ip$, respectively. Then the differential equation takes the form

$$\frac{\partial}{\partial t} W(x, p, t) = -\left(\frac{p}{m} \right) \frac{\partial}{\partial x} W(x, p, t)$$

$$+ \sum_{n=0}^{\infty} \left(\frac{-i}{2} \right)^{2n} \frac{1}{(2n + 1)!} \left[\left(\frac{\partial}{\partial x} \right)^{2n+1} V(x) \right] \left[\left(\frac{\partial}{\partial p} \right)^{2n+1} W(x, p, t) \right]. \tag{5.19}$$

If $V(x)$ does not take the form of a finite polynomial this is an infinite-order differential equation. However, if

$$\left(\frac{\partial}{\partial x} \right)^3 V(x) = 0, \tag{5.20}$$

then the above equation can be reduced to the classical Liouville equation:

$$\frac{\partial}{\partial t} W(x, p, t) = \left(\frac{\partial H}{\partial x} \right) \frac{\partial}{\partial p} W(x, p, t) - \left(\frac{\partial H}{\partial p} \right) \frac{\partial}{\partial x} W(x, p, t). \tag{5.21}$$

If the potential is of the form $V(x) = ax^2 + bx + c$, we see that the Wigner distribution function satisfies the Liouville equation. There is a clear difference between them even though the same mathematics can be used for both the classical phase-space distribution and the Wigner distribution function. Classically the distribution is a probability distribution in phase space [3, 4], while the Wigner distribution is not a probability distribution in phase space [12].

For a free particle with $V = 0$, the differential equation of equation (5.19) becomes

$$\frac{\partial}{\partial t} W(x, p, t) = -\left(\frac{p}{m} \right) \frac{\partial}{\partial x} W(x, p, t). \tag{5.22}$$

This equation will be used in section 5.3 to study wave packet spread.

If we consider the harmonic oscillator where $V(x) = Kx^2/2$, the differential equation is

$$\frac{\partial}{\partial t}W(x,\,p,\,t) = -\left(\frac{p}{m}\right)\frac{\partial}{\partial x}W(x,\,p,\,t) + (Kx)\frac{\partial}{\partial p}W(x,\,p,\,t). \tag{5.23}$$

We shall study this equation in detail in section 5.4.

When the potential is of the form $V(x) = gx$, as it is when the particle is in a constant gravitational field, the quantum Liouville equation is [12]

$$\frac{\partial}{\partial t}W(x,\,p,\,t) = -\left(\frac{p}{m}\right)\frac{\partial}{\partial x}W(x,\,p,\,t) + g\frac{\partial}{\partial p}W(x,\,p,\,t). \tag{5.24}$$

The solution of this differential equation is

$$W(x,\,p,\,t) = W(x - pt/m,\,p + gt,\,t = 0), \tag{5.25}$$

when the Wigner distribution function is known at $t = 0$. As time progresses, $x = 0$ moves to pt/m while the phase-space distribution moves from $p = 0$ to $p = -gt$. The solution can therefore be represented by a coordinate transformation,

$$\begin{pmatrix} x' \\ p' \\ 1 \end{pmatrix} = \begin{pmatrix} 1 & t/m & 0 \\ 0 & 1 & -gt \\ 0 & 0 & 1 \end{pmatrix}\begin{pmatrix} x \\ p \\ 1 \end{pmatrix}, \tag{5.26}$$

which means that the solution of the Liouville equation for the linear potential can be represented by a linear coordinate transformation. By using the coordinate transformation in phase space we can learn many interesting lessons in physics as will be seen in later chapters as well as in sections 5.3 and 5.4.

5.3 Wave packet spread

Caused by the fact that particles have different phase velocities for different momenta, the wave packet spread for a free particle is one of the most puzzling features of the Schrödinger picture of quantum mechanics. Because of this difference in phase velocities for different momenta, the position distribution becomes widespread while the momentum distribution remains invariant. The uncertainty product $\langle \Delta x \rangle \langle \Delta p \rangle$ thus increases as time progresses or regresses. We then are prompted to ask whether there is a way of stating this uncertainty relation in a time-independent manner.

For a free particle the differential equation given in equation (5.22) is satisfied by the Wigner function and its solution is [10, 13–15, 19]

$$W(x,\,p,\,t) = W(x - pt/m,\,p,\,0). \tag{5.27}$$

For a Wigner function localized within a given region in phase space at $t = 0$, its area remains invariant, while its localization region is sheared.

Since both x and p are c-numbers in the phase-space picture of quantum mechanics and $(\Delta x)(\Delta p)$ cannot be smaller than Planck's constant, the uncertainty relation can be quantified in terms of the area within which the Wigner function is concentrated. Thus the localization area cannot be smaller than Planck's constant. Although this region may undergo deformation, the uncertainty remains invariant.

This is true as long as the area of localization is invariant under certain transformations. This region of localization is called the error box [1].

Let us consider, for example, the spread of a Gaussian wave packet. Its momentum distribution is given by

$$g(p) = \left(\frac{b}{\pi}\right)^{1/4} e^{-bp^2/2} \tag{5.28}$$

at $t = 0$. The momentum distribution function, as time progresses, remains unchanged:

$$|g(p)|^2 = \left(\frac{b}{\pi}\right)^{1/2} e^{-bp^2}. \tag{5.29}$$

Defining (Δp) as

$$(\Delta p) = \left\{ 2 \int |g(p)|^2 p^2 \, dp \right\}^{1/2} \tag{5.30}$$

leads to $(\Delta p) = (1/b)^{1/2}$, which, under time evolution, is invariant.

By constructing

$$\psi(x, t) = \left(\frac{1}{2\pi}\right)^{1/2} \int e^{(ipx - ip^2 t/2\,m)} g(p) \, dp, \tag{5.31}$$

the time-dependent Schrödinger wave function becomes

$$\psi(x, t) = \left(\frac{b}{\pi}\right)^{1/4} \left(\frac{1}{b + it/m}\right)^{1/2} e^{-x^2/2(b + it/m)}. \tag{5.32}$$

If we consider that the spatial probability distribution is

$$|\psi(x, t)|^2 = \left\{ b/\pi(b^2 + (t/m)^2) \right\}^{1/2} e^{-bx^2/(b^2 + (t/m)^2)}, \tag{5.33}$$

we obtain

$$(\Delta x) = \left\{ b(1 + (t/mb)^2) \right\}^{1/2}. \tag{5.34}$$

In the Schrödinger picture the uncertainty product becomes

$$(\Delta x)(\Delta p) = \left(1 + (t/mb)^2 \right)^{1/2}, \tag{5.35}$$

which increases as t progresses. In Schrödinger's form of quantum mechanics this is widely known as wave packet spread.

Now the question of whether, in the phase-space picture of quantum mechanics, the uncertainty can be stated in a time-invariant manner becomes of interest. The Wigner function for the above wave function can be constructed as

$$W(x, p, t) = \frac{1}{\pi} e^{-(x - pt/m)^2/(b + bp^2)}. \tag{5.36}$$

Thus, this distribution is concentrated within the region where the exponent is less than one in magnitude:

$$(x - pt/m)^2/b + bp^2 < 1. \tag{5.37}$$

It is possible to describe this region by the tilted ellipse in figure 5.1, which at $t = 0$ is localized within the elliptic region $x^2/b + bp^2 = 1$. As time progresses, the elliptic region becomes sheared.

Thus the magnitude of the uncertainty can be defined as the area divided by π. Because the area remains unchanged as time progresses, the uncertainty remains invariant. The volume of the error box is the area which becomes sheared as time progresses. However, its volume remains invariant.

We can once again describe the time evolution in terms of a linear transformation in phase space:

$$\begin{pmatrix} x' \\ p' \end{pmatrix} = \begin{pmatrix} 1 & t/m \\ 0 & 1 \end{pmatrix} \begin{pmatrix} x \\ p \end{pmatrix}. \tag{5.38}$$

The mathematical property of this transformation has been discussed in [12].

5.4 Harmonic oscillators

The one-dimensional harmonic oscillator can be described in terms of the classical Hamiltonian as

$$H = \frac{1}{2m}p^2 + \frac{K}{2}x^2. \tag{5.39}$$

Then p and x can be measured in units of $\sqrt{m}$ and $1/\sqrt{K}$, respectively. Consequently we can write the Hamiltonian as

$$H = \frac{1}{2}(p^2 + x^2). \tag{5.40}$$

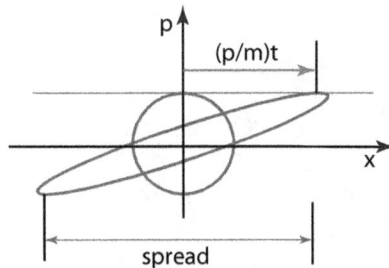

Figure 5.1. The spread of the Gaussian wave packet. As time progresses or regresses, the uncertainty product $(\Delta x)(\Delta p)$ increases in the Schrödinger picture. The area of localization is Planck's constant, which remains invariant [12].

The Schrödinger equation then becomes, in this system of units,

$$i\frac{\partial}{\partial t}\psi(x,\ t) = -\left(\frac{1}{2}\right)\left(\frac{\partial}{\partial x}\right)^2\psi(x,\ t) + \left(\frac{1}{2}x^2\right)\psi(x,\ t).$$
(5.41)

For the Wigner function the Liouville equation then takes the form

$$\frac{\partial}{\partial t}W(x,\ p,\ t) = \left(x\frac{\partial}{\partial p} - p\frac{\partial}{\partial x}\right)W(x,\ p,\ t),$$
(5.42)

with solution

$$W(x,\ p,\ t) = \exp\left\{\left(x\frac{\partial}{\partial p} - p\frac{\partial}{\partial x}\right)t\right\}W(x,\ p,\ 0).$$
(5.43)

In phase space this represents a rotation around the origin, which in matrix form is given by

$$\begin{pmatrix} x' \\ p' \end{pmatrix} = \begin{pmatrix} \cos(t) & -\sin(t) \\ \sin(t) & \cos(t) \end{pmatrix}\begin{pmatrix} x \\ p \end{pmatrix}.$$
(5.44)

When the distribution is invariant under rotations, this is independent of time. We can write the time-independent Schrödinger equation as

$$-\left(\frac{1}{2}\right)\left(\frac{\partial}{\partial x}\right)^2\psi(x) + \left(\frac{1}{2}x^2\right)\psi(x) = (n + 1/2)\psi(x),$$
(5.45)

and then the normalized solutions have the form

$$\psi_n(x) = \left[\frac{1}{\sqrt{\pi}\,2^n n!}\right]^{1/2} H_n(x)e^{-x^2/2}.$$
(5.46)

Here $H_n(x)$ is the Hermite polynomial of the nth order. Since the wave function is in an energy eigenstate, the Wigner function can be evaluated as

$$W_n(x,\ p) = \frac{1}{\pi} \int \psi_n^*(x + y)\psi_n(x - y)e^{2ipy}dy.$$
(5.47)

The resulting calculation gives

$$W_n(x,\ p) = \left(\frac{n!}{\pi}\right)(e^{(-r^2/2)}) \sum_{k=0}^{\infty}(-1)^k r^{2(n-k)}/([(n - k)!]^2 k!),$$
(5.48)

where

$$r^2 = 2(x^2 + p^2).$$

Thus $W_n(x,\ p)$ is a function only of r, and can be written as $W_n(r)$. This satisfies the differential equation [9]

$$-\frac{1}{2r}\left[\frac{d}{d\rho}r\left(\frac{d}{dr}W_n(r)\right)\right] + \frac{1}{2}r^2 W_n(r) = (2n+1)W_n(r), \tag{5.49}$$

the solution to which is readily available in the literature [2, 12, 21]. Its form is

$$W_n(x,p) = ((-1)^n/\pi)(L_n(r^2))e^{-r^2/2}, \tag{5.50}$$

where $L_n(r^2)$ is the Laguerre polynomial. Since this expression is invariant under rotations around the origin, it can also be written as [21]

$$W_n(x,p) = \frac{1}{\pi}\left(\frac{1}{4}\right)^n e^{-(x^2+p^2)} \sum_{k=0}^{n}\left(\frac{1}{k!(n-k)!}\right)H_{2k}(\sqrt{2}x)H_{2(n-k)}(\sqrt{2}p). \tag{5.51}$$

In this equation, $H_{2k}(\sqrt{2}x)$ and $H_{2(n-k)}(\sqrt{2}p)$ are the Hermite polynomials.

If the Wigner function is in the ground state then

$$W_0(x,p) = \frac{1}{\pi}e^{-(x^2+p^2)}. \tag{5.52}$$

When $n = 1$ the Wigner function becomes

$$W_1(x,p) = \frac{2}{\pi}\left(x^2 + p^2 - \frac{1}{2}\right)e^{-(x^2+p^2)}. \tag{5.53}$$

Although $W_0(x,p)$ is positive everywhere in phase space, as is illustrated in figure 5.2, $W_1(x,p)$ is negative at the origin, but is positive for sufficiently large values of $(x^2 + p^2)$. For very large values of $(x^2 + p^2)$ both of them become vanishingly small. We should note that the probability density in x is always positive in spite of the fact that $W_1(x,p)$ is negative around the origin.

5.5 Minimum uncertainty in phase space

The harmonic oscillator in the ground state, with Gaussian space and momentum wave functions produces the minimum uncertainty distribution. This aspect is shown clearly in the Wigner phase space. Let us rewrite the Wigner function as

$$W_0 = e^{-(x^2+p^2)}. \tag{5.54}$$

We can describe this function as a circle

$$x^2 + p^2 = 1. \tag{5.55}$$

This is known as the minimum area allowed by the present form of quantum mechanics. We can translate the position of this circle along the x and p directions. Since these translations do not interfere with each other, they do not cause any mathematical complications. This aspect of translations is illustrated in figure 5.3.

In addition, this circle can be squeezed to

$$e^{-2\eta}x^2 + e^{2\eta}p^2 = 1. \tag{5.56}$$

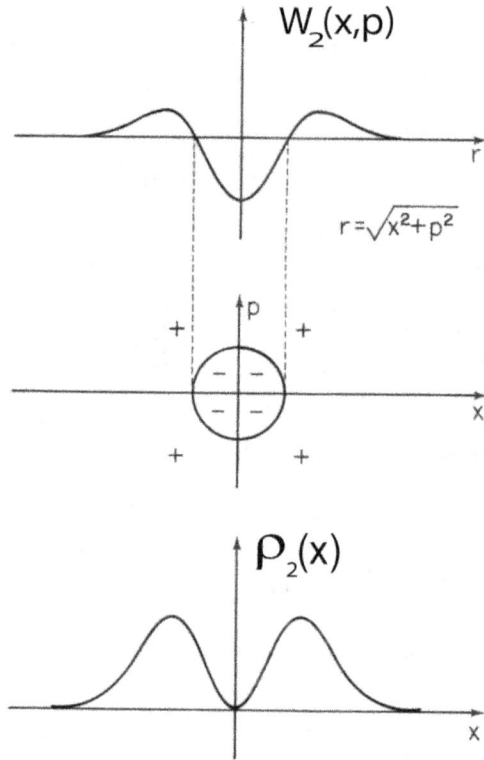

$W_2(x,p)$

$r=\sqrt{x^2+p^2}$

$P_2(x)$

Figure 5.2. The Wigner function for the harmonic oscillator in the first excited state, which is negative at the origin, becomes positive as $(x^2 + p^2)$ increases, and becomes vanishingly small as $(x^2 + p^2)$ becomes very large. When the Wigner function is integrated over p, it becomes the probability density in x, which is positive for all values of x [12].

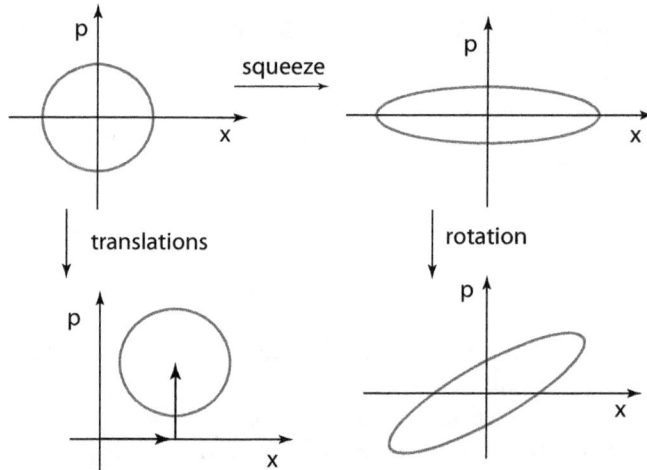

squeeze

translations

rotation

Figure 5.3. The coordinate translations of this minimum uncertainty circle do not change the area. The circle can also be squeezed and rotated. They are all uncertainty-preserving transformations. In figure 5.1 the circle is sheared. This is a special case of a squeeze followed by a rotation.

This is an area-preserving transformation. In terms of matrices, this transformation can be written as

$$\begin{pmatrix} e^{\eta} & 0 \\ 0 & e^{-\eta} \end{pmatrix}. \tag{5.57}$$

This ellipse can then be rotated with the transformation matrix

$$\begin{pmatrix} \cos\theta & -\sin\theta \\ \sin\theta & \cos\theta \end{pmatrix}. \tag{5.58}$$

The transformation matrices given in equations (5.57) and (5.58) can now be written as

$$\begin{pmatrix} e^{\eta} & 0 \\ 0 & e^{-\eta} \end{pmatrix} = e^{2i\eta K_3} \quad \text{and} \quad \begin{pmatrix} \cos\theta & -\sin\theta \\ \sin\theta & \cos\theta \end{pmatrix} = e^{2i\theta J_2}, \tag{5.59}$$

with

$$K_3 = \frac{1}{2}\begin{pmatrix} i & 0 \\ 0 & -i \end{pmatrix} \quad \text{and} \quad J_2 = \frac{1}{2}\begin{pmatrix} 0 & -i \\ i & 0 \end{pmatrix}. \tag{5.60}$$

The commutation of these two matrices leads to another matrix:

$$[J_2, K_3] = iK_1, \tag{5.61}$$

with

$$K_1 = \frac{1}{2}\begin{pmatrix} 0 & i \\ i & 0 \end{pmatrix}, \tag{5.62}$$

which leads to

$$e^{2i\lambda K_1} = \begin{pmatrix} \cosh\lambda & \sinh\lambda \\ \sinh\lambda & \cosh\lambda \end{pmatrix}. \tag{5.63}$$

This K_1 matrix can have commutation relations with K_3 and J_2:

$$[K_1, K_3] = iJ_2 \quad \text{and} \quad [K_1, J_2] = iK_3. \tag{5.64}$$

As noted in section 2.5, the three matrices K_1, K_3, and J_2 form a closed set of the commutation relations.

The group of transformations generated by these three matrices is called the two-dimensional symplectic or $Sp(2)$ group [6]. We are of course familiar with the group of two-by-two matrices corresponding to the three-dimensional rotation group. Unlike the rotation-like matrices generated by the three Pauli matrices, the transformation matrices of this symplectic group are all real and can be graphically described in the two-dimensional space, as illustrated in figure 5.3. As we shall see in later chapters, this $Sp(2)$ group serves as one of the basic scientific languages in optical sciences.

5.6 Density matrix

We noted in section 5.1 that the density matrix provides the transition from the Schrödinger wave function to the Wigner function. We examine in this section how some of the basic properties of the density matrix can be translated into the language of the Wigner function. Recall that for one pair of canonical variables, the Wigner function can be written as

$$W(x, p) = \frac{1}{\pi} \int \rho(x + y, x - y)e^{2ipy}dy. \tag{5.65}$$

It is possible, from this, to convert the Wigner function back to the density matrix:

$$\rho(x, x') = \int W\left(\frac{x + x'}{2}, p\right)e^{-ip(x-x')}dp. \tag{5.66}$$

With the help of this relation, we can calculate from the Wigner function $\mathrm{Tr}(\rho)$ and $\mathrm{Tr}(\rho^2)$.

Hence, $\mathrm{Tr}(\rho)$ can be written as

$$\mathrm{Tr}(\rho) = \int \rho(x, x)dx = \int W(x, p)dxdp = 1. \tag{5.67}$$

For two different density matrices ρ_1 and ρ_2, we can construct two different Wigner functions W_1 and W_2 as

$$\begin{aligned} Tr(\rho_1\rho_2) &= \int \left\{ \int \rho_1(x, x')\rho_2(x', x)dx' \right\} dx \\ &= 2\pi \int W_1(x, p)W_2(x, p)dxdp. \end{aligned} \tag{5.68}$$

This leads to

$$\begin{aligned} Tr(\rho^2) &= \int \rho(x, x')\rho(x', x)dxdx' \\ &= 2\pi \int (W(x, p))^2 dxdp. \end{aligned} \tag{5.69}$$

For a more concrete example, we can consider a one-dimensional harmonic oscillator in thermal equilibrium. We can then start from the density matrix of equation (1.97) where k_B designates Boltzmann's constant. The density matrix for an oscillator with unit frequency, takes the form

$$\rho_T(x, x') = (1 - e^{-1/k_BT}) \sum_n e^{-n/k_BT}\psi_n(x)\psi_n^*(x'). \tag{5.70}$$

Using the analysis given in section 5.4, the Wigner function for this system is

$$W_T(x, p) = (1 - e^{-1/k_BT}) \sum_n e^{-n/k_BT} W_n(x, p), \tag{5.71}$$

where $W_n(x, p)$ is the Wigner function for the nth excited-state harmonic oscillator.

The series expansion of a function in terms of unfamiliar functions is not very useful. We thus ask the question of whether a closed form is available for $W_T(x, p)$ or $\rho_T(x, x')$. For this purpose, we start with the trial function [11, 20]:

$$f(x, x') = \left(\frac{1}{\pi}\right)^{1/2} \exp\left\{-\left(\frac{1}{4}\right)\left(e^{-2\eta}(x + x')^2 + e^{2\eta}(x - x')^2\right)\right\}, \tag{5.72}$$

and then expand the expression in terms of the harmonic oscillator wave functions:

$$f(x, x') = \sum_n C_{nm}\psi_n(x)\psi_m(x'). \tag{5.73}$$

The coefficient C_{nm} is

$$C_{nm} = \int \psi_n^*(x)\psi_m^*(x')f(x, x')dxdx'. \tag{5.74}$$

Evaluating this integral by using the generating function for the Hermite polynomials [16] gives

$$e^{(2sx-s^2)} = \sum_n \frac{s^2}{n!}H_n(x). \tag{5.75}$$

Thus in terms of the harmonic oscillator wave functions, this generating function takes the form

$$\left(\frac{1}{\pi}\right)^{1/4} e^{(2sx-s^2-x^2/2)} = \sum_n s^n(2^n/n!)^{1/2}\psi_n(x). \tag{5.76}$$

Hence

$$\left(\frac{1}{\pi}\right)^{1/2} \int f(x, x')e^{(2sx+2rx'-s^2-r^2-(x^2+x'^2)/2)}dxdx'$$
$$= \sum_m \sum_n s^n r^m (2^n 2^m/(n!m!))^{1/2} C_{nm}. \tag{5.77}$$

The integral can be evaluated yielding

$$(\sqrt{\pi}/\cosh\eta)e^{(2rs(\tanh\eta))} = \sum_m \sum_n s^n r^m (2^n 2^m/(n!m!))^{1/2} C_{nm}. \tag{5.78}$$

The left-hand side of the power series of $(2rs)$ can now be expanded to obtain

$$C_{nm} = (\sqrt{\pi}/\cosh\eta)(\tanh\eta)^n \delta_{nm}. \tag{5.79}$$

This leads to

$$\rho_T(x, x') = \left(\frac{\tanh(1/2k_BT)}{\pi}\right)^{1/2} \exp\left\{-\left(\frac{1}{4}\right)((x + x')^2 \tanh(1/2k_BT)\right.$$

$$\left. + (x - x')^2 \coth(1/2k_BT))\right\}. \tag{5.80}$$

From this version of the density matrix we can reproduce $\text{Tr}(\rho) = 1$ and $\text{Tr}(\rho^2) = \tanh(1/2k_BT)$ given in equations (1.98) and (1.100), which were derived from the series form of $\rho_T(x, x')$. This is a pure state in the zero-temperature limit.

The above expression for $\rho_T(x, x')$, allows us to derive

$$W_T(x, p) = \left(\frac{\tanh(1/2k_BT)}{\pi}\right) \exp\left\{-(x^2 + p^2)\tanh(1/2k_BT)\right\}. \tag{5.81}$$

When T approaches zero, this expression becomes that of the ground-state harmonic oscillator. Additionally, the traces of the density matrix can then be calculated as

$$\text{Tr}(\rho_T) = \int W_T(x, p)dxdp = 1 \tag{5.82}$$

and

$$\text{Tr}(\rho_T\rho_T) = 2\pi \int (W_T(x, p))^2 dxdp = \tanh(1/2k_BT). \tag{5.83}$$

5.7 Measurable quantities

From transition probabilities and expectation values we can calculate measurable quantities when using the Schrödinger picture of quantum mechanics. It should also be possible to calculate these quantities directly from the Wigner function as this function is constructed from the Schrödinger wave function.

To study the transition probability, we consider first two different wave functions $\psi(x)$ and $\phi(x)$, and their corresponding Wigner function $W_\psi(x, p)$ and $W_\phi(x, p)$. Using these Wigner functions the transition probability takes the form of equation (5.12). Now we can rewrite equation (5.12) as

$$|(\phi(x), \psi(x))|^2 = 2\pi \int W_\psi(x, p)W_\phi(x, p)dxdp. \tag{5.84}$$

If both of the Wigner functions in the integrand are constructed from the pure states of $\phi(x)$ and $\psi(x)$, then this expression is valid. However, when the states are non-pure the density matrix is the appropriate quantity. If we rewrite equation (5.68) as

$$\text{Tr}\,(\rho_1\rho_2) = 2\pi \int W_1(x, p) W_2(x, p) dx dp, \tag{5.85}$$

then the Wigner function accommodates both pure and non-pure states, and equation (5.69) is a special case of equation (5.68).

For instance from equation (5.51) we know that $W_0(x, p)$ and $W_1(x, p)$ are pure-state Wigner functions for the ground and first-excited oscillator states, respectively. Then it can be verified by direct calculation that

$$2\pi \int (W_0(x, p))^2 dx dp = 2\pi \int (W_1(x, p))^2 dx dp = 1 \tag{5.86}$$

and

$$2\pi \int W_0(x, p) W_1(x, p) dx dp = 0. \tag{5.87}$$

Additionally, when n and m are different, if we use the differential equation of (5.24), it can be shown that the orthogonality relation

$$\int W_n(x, p) W_m(x, p) dx dp = 0 \tag{5.88}$$

holds.

Furthermore, from $W_T(x, p)$ of equation (5.71), which is the Wigner function for a thermally excited state, we know that this is a non-pure state. When the Wigner function is in the nth excited state we can write the probability of W_T as

$$P_n = 2\pi \int W_T(x, p) W_n(x, p) dx dp = (-1)^n \tanh\left(\frac{1}{2k_B T}\right)$$
$$\times \int_0^\infty L_n(\rho^2) \exp\left\{-(\rho^2/2)\left(1 + \tanh\left(\frac{1}{2k_B T}\right)\right)\right\}(2\rho) d\rho. \tag{5.89}$$

From the formula [5]

$$\int_0^\infty e^{-bx} L_n(x) dx = \frac{1}{b}\left(1 - \frac{1}{b}\right)^n, \tag{5.90}$$

we can evaluate this integral. This calculation results in

$$P_n = (1 - e^{-1/k_B T}) e^{-n/k_B T}, \tag{5.91}$$

as would be expected from the definition of $\rho(x, x')$.

We would like now to consider the expectation value of an operator applicable to $\psi(x)$ or to $\chi(p)$, which would apply to both pure and non-pure states. To do so, we use the operator Q which has the form

$$Q = A(x) + B(p), \tag{5.92}$$

where $A(x)$ and $B(p)$ depend only on x and p, respectively. Then we obtain the expectation value

$$\langle Q \rangle = (\psi(x), \hat{A}(x)\psi(x)) + (\chi(p), \hat{B}(p)\chi(p))$$
$$= \int (A(x) + B(p)) W(x, p) dx dp, \tag{5.93}$$

where $\hat{A}$ and $\hat{B}$ are the q-number operators in the Schrödinger picture. It is now clear that we are using the term *expectation value* for both pure and non-pure states and that the above formula is valid for both cases.

Let us now consider the energy operator for the harmonic oscillator:

$$\hat{H} = \frac{1}{2}\left(x^2 - \left(\frac{\partial}{\partial x}\right)^2\right). \tag{5.94}$$

The expectation value for this is given by

$$\langle H \rangle = \frac{1}{2}\left\{\left(\psi(x), x^2\psi(x)\right) + \left(\chi(p), p^2\chi(p)\right)\right\}$$
$$= \frac{1}{2}\int (x^2 + p^2) W(x, p) dx dp. \tag{5.95}$$

Using the Wigner function for the nth excited-state oscillator, $\langle H \rangle$ should result in $(n + 1/2)$. We obtain for the thermally excited state,

$$\langle H \rangle = \frac{1}{2}\int (x^2 + p^2) W_T(x, p) dx dp$$
$$= \frac{1}{2}\left\{\tanh\left(\frac{1}{2k_B T}\right)\right\}, \tag{5.96}$$

which is expected as well from the density matrix.

Since the operator Q depends on both x and p one can use the phase-space picture to show in a straight-forward manner that [7, 8, 12]

$$\int xp W(x, p) dx dp = -\left(\frac{i}{2}\right)\left(\psi(x), \left(x\frac{\partial}{\partial x} + \frac{\partial}{\partial x}x\right)\psi(x)\right) \tag{5.97}$$

and

$$\int x^2 p^2 W(x, p) dx dp$$
$$= -\left(\frac{1}{4}\right)\left(\psi(x), \left(x^2\left(\frac{\partial}{\partial x}\right)^2 + 2x\left(\frac{\partial}{\partial x}\right)^2 x + \left(\frac{\partial}{\partial x}\right)^2 x^2\right)\psi(x)\right). \tag{5.98}$$

The above equations represent special cases of a more general relation:

$$\int W(x, p)(x^n p^m) dx dp = (-i)^m \left(\frac{1}{2}\right)^n \sum_{r=0}^{n} \binom{n}{r}\left(\psi(x), x^{n-r}\left(\frac{\partial}{\partial x}\right)^m x^r \psi(x)\right). \tag{5.99}$$

We can derive this from the theorem stating

$$\left(\psi(x),\ e^{\left(-i\sigma x-\tau\frac{\partial}{\partial x}\right)}\psi(x)\right) = \int W(x, p)e^{-i(\sigma x+\tau p)}dxdp, \tag{5.100}$$

where σ and τ are scalar parameters. Then equation (5.99) can be derived by expanding the exponential functions of both sides and by choosing the coefficients of $\sigma^n\tau^m$.

By using the Baker–Campbell–Hausdorff formula [17],

$$e^{\left(-i\sigma x-\tau\frac{\partial}{\partial x}\right)} = e^{(i\sigma\tau/2)}e^{(-i\sigma x)}e^{\left(-\tau\frac{\partial}{\partial x}\right)}, \tag{5.101}$$

it is possible to prove the above theorem [9, 12, 18]. The left-hand side of equation (5.100) therefore becomes

$$e^{i\sigma\tau/2}(\psi(x),\ e^{-i\sigma x}\psi(x - \tau)). \tag{5.102}$$

If we change the variable y to $y/2$, the Wigner function given in equation (5.6) can be written as

$$W(x, p) = \frac{1}{2\pi} \int \psi^*(x + y/2)\psi(x - y/2)e^{ipy}dy. \tag{5.103}$$

Hence the right-hand side of equation (5.100) has the form

$$\frac{1}{2\pi} \int \psi^*(x + y/2)\psi(x - y/2)e^{ip(y-\tau)}e^{i\sigma x}dxdydp, \tag{5.104}$$

which when integrated over p and y leads to

$$\int \psi^*(x + \tau/2)\psi(x - \tau/2)e^{-i\sigma x}dx. \tag{5.105}$$

Similarly changing the variable x to $x - \tau/2$, gives the form

$$e^{i\sigma\tau/2} \int \psi^*(x)\psi(x - \tau)e^{-i\sigma x}dx, \tag{5.106}$$

identical to the expression of equation (5.102), which completes the proof.

It is now possible to generalize equation (5.100) to

$$\int A(\sigma, \tau)\left\{\int W(x, p)e^{-i(\sigma x+\tau p)}dxdp\right\}d\sigma d\tau$$
$$= \int A(\sigma, \tau)\left[\psi(x),\ e^{-\left(i\sigma x+\tau\frac{\partial}{\partial x}\right)}\psi(x)\right]d\sigma d\tau. \tag{5.107}$$

The c-number operator

$$Q(x, p) = \int A(\sigma, \tau)e^{-i(\sigma x+\tau p)}d\sigma d\tau \tag{5.108}$$

corresponds now to the q-number operator

$$\hat{Q}(x, p) = \int A(\sigma, \tau)e^{-\left(i\sigma x+\tau\frac{\partial}{\partial x}\right)}d\sigma d\tau \tag{5.109}$$

in the Schrödinger picture.

References

[1] Caves C M 1981 Quantum-mechanical noise in an interferometer *Phys. Rev.* D **23** 1693–708
[2] Davies R W, and Davies K T R 1975 On the Wigner distribution function for an oscillator *Ann. Phys.* **89** 261–73
[3] Feynman R P 1998 *Statistical Mechanics: A Set of Lectures* (*Advanced Book Classics*) (Boulder, CO: Westview)
[4] Goldstein H 1980 *Classical Mechanics* (*Addison-Wesley Series in Physics*) 2nd edn (Reading, MA: Addison-Wesley)
[5] Gradshteïn I S, Ryzhik I M, and Jeffrey A 2007 *Table of Integrals, Series and Products* 7th edn (Amsterdam: Academic)
[6] Guillemin V and Sternberg S 2001 *Symplectic Techniques in Physics* (Cambridge: Cambridge University Press)
[7] Han D, Kim Y S, and Noz M E 1988 Linear canonical transformations of coherent and squeezed states in the Wigner phase space *Phys. Rev.* A **37** 807–14
[8] Han D, Kim Y S, and Noz M E 1989 Linear canonical transformations of coherent and squeezed states in the Wigner phase space. II. Quantitative analysis *Phys. Rev.* A **40** 902–12
[9] Hillery M, O'Connell R F, Scully M O, and Wigner E P 1984 Distribution functions in physics: fundamentals *Phys. Rep.* **106** 121–67
[10] Kijowski J 1974 On the time operator in quantum mechanics and the Heisenberg uncertainty relation for energy and time *Rep. Math. Phys.* **6** 361–86
[11] Kim Y S and Noz M E 1986 *Theory and Applications of the Poincaré Group* (Dordrecht: Springer)
[12] Kim Y S and Noz M E 1991 *Phase Space Picture of Quantum Mechanics: Group Theoretical Approach* (*Lecture Notes in Physics* vol 40) (Singapore: World Scientific)
[13] Kim Y S and Wigner E P 1990 Canonical transformation in quantum mechanics *Am. J. Phys* **58** 439–48
[14] Lee H-W and Scully M O 1982 Wigner phase-space description of a Morse oscillator *J. Chem. Phys.* **77** 4604–10
[15] Littlejohn R 1986 The semiclassical evolution of wave packets *Phys. Rep.* **138** 193–291
[16] Magnus W, Oberhettinger F, and Soni R P 1966 *Formulas and Theorems for the Special Functions of Mathematical Physics* (Berlin: Springer)
[17] Miller W 1972 *Symmetry Groups and Their Applications Pure and Applied Mathematics* vol 50 (New York: Academic)
[18] Moyal J E 1949 Quantum mechanics as a statistical theory *Math. Proc. Camb. Philos. Soc.* **45** 99–124
[19] Royer A 1985 Cumulant approximations and renormalized Wigner–Kirkwood expansion for quantum Boltzmann densities *Phys. Rev.* A **32** 1729–43

[20] Ruiz M J 1974 Orthogonality relation for covariant harmonic-oscillator wave functions *Phys. Rev.* D **10** 4306–7

[21] Schleich W, Walls D F, and Wheeler J A 1988 Area of overlap and interference in phase space versus Wigner pseudoprobabilities *Phys. Rev.* A **38** 1177–86

[22] Wigner E P 1932 On the quantum correction for thermodynamic equilibrium *Phys. Rev.* **40** 749–59

IOP Publishing

Mathematical Devices for Optical Sciences

Sibel Başkal, Young S Kim, and Marilyn E Noz

Chapter 6

Coherent states of light

The harmonic oscillator played a pivotal role during the development of quantum mechanics. The most controversial aspect of quantum mechanics during its developing stage was its uncertainty relation. The problem to be solved was how to minimize the uncertainty product, and it was shown that the Gaussian wave function produces this desired product.

This Gaussian form is of course the ground-state solution of the Schrödinger equation. However, the uncertainty product becomes larger for excited states. Then, the question arose as to whether it would be possible to excite the oscillator while preserving the minimum-uncertainty product.

For the single oscillator system it is not possible. On the other hand, if all excited states are linearly superposed in a particular manner, then the result preserves the minimum-uncertainty product [16]. This linear combination is called the coherent state [2, 7, 12, 17].

However, its physical role was not appreciated until the harmonic oscillator wave functions became useful for the photon number state [4, 6]. The nth excited state corresponds to the state of n photons. Thus, the coherent harmonic oscillator state can serve as the coherent superposition of the photon number states. This coherent photon produces a Poisson distribution of the probability in the photon number, which was actually observed in laser cavities [1].

In this chapter, we shall discuss mathematical issues in studying the coherent state of light. As we noted in chapter 1 on the forms of quantum mechanics, the harmonic oscillator plays a central role in quantum mechanics and also in quantum field theory based on the Fock space [6].

We start from the harmonic oscillator differential equation of (1.133), with its solution $\chi_n(x)$ as given in equation (1.134). This wave function is for the nth excited state. As is well known, this number n for the excited state serves as the number of particles in quantum field theory. Thus, instead of the wave function, we use the ket vector $|n\rangle$. Number states are orthonormal and this condition is expressed as

$$\langle n'|n\rangle = \delta_{nn'}. \tag{6.1}$$

As we did in equation (1.132) we define the operators

$$a = \frac{1}{\sqrt{2}}\left(x + \frac{\partial}{\partial x}\right) \quad \text{and} \quad a^\dagger = \frac{1}{\sqrt{2}}\left(x - \frac{\partial}{\partial x}\right). \tag{6.2}$$

Then

$$a|n\rangle = \sqrt{n}|n-1\rangle \quad \text{and} \quad a^\dagger|n\rangle = \sqrt{n+1}|n+1\rangle. \tag{6.3}$$

Since the operators a and $a^\dagger$ reduce and increase the number of particles by one, they are called the annihilation and creation operators, respectively.

Next, we introduce the operator

$$N = a^\dagger a, \tag{6.4}$$

which results in

$$N|n\rangle = n|n\rangle. \tag{6.5}$$

Thus, N is the number operator which measures the number of particles for a given ket vector. The n-photon state $|n\rangle$ can thus be written as

$$|n\rangle = \frac{1}{\sqrt{n!}}(a^\dagger)^n |0\rangle, \tag{6.6}$$

where $|0\rangle$ is the zero-photon or vacuum state.

In section 6.1 we elaborate on this number operator as a physically measurable variable and on its uncertainty relation with the phase of the wave. In section 6.2 it is noted that the exponentiation of the sum of two operators is not necessarily the product of the separate exponentials of those two operators. The mathematical tool needed to deal with this problem is called the Baker–Campbell–Hausdorff formula, often abbreviated as the BCH formula. In this section, we shall study this mathematics at length.

In section 6.3 we shall study the coherent state of light in detail. It will be shown that the coherent state is the eigenstate of the annihilation operator,

$$a|\alpha\rangle = \alpha|\alpha\rangle, \tag{6.7}$$

where α is a complex variable. It will be shown that $|\alpha\rangle$ is the minimum-uncertainty state. We shall also discuss how this state manifests itself in the real world of quantum optics

In section 6.4 it is noted that the complex number α consists of two real numbers. Thus, the symmetry can be studied in a two-dimensional plane. It is shown that the coherent state can be constructed from the vacuum state $|0\rangle$ by translations and rotations in the two-dimensional complex plane of the parameter α. However, it is pointed out that these translations and rotations do not constitute the two-dimensional Euclidean group.

In section 6.5 the symmetry problem presented in the previous section is studied further within the frame work of the Wigner phase-space picture, and it is noted that the symmetry in the complex α-plane is indeed that of the two-dimensional Euclidean group.

In section 6.6 it is shown that the squeeze operation is also possible in the complex α plane. It is shown that this squeeze operation could address the issue of the c-number number–phase uncertainty relation.

6.1 Phase-number uncertainty relation

The concept of free photons is well established in terms of their creation and annihilation operators. These operations can be defined in terms of the mathematics describing the energy levels of the harmonic oscillator. The oscillator in this case is not a mechanical oscillator, but is defined in the space commonly known as the Fock space [6], where a and $a^{\dagger}$, instead of x and p, are used as the basic dynamical variables.

However, according to the expressions given for a and $a^{\dagger}$ in equation (6.2), we are led to write the uncertainty principle in the Fock space as

$$[a, a^{\dagger}] = 1, \quad [a, a] = 0, \quad \text{and} \quad [a^{\dagger}, a^{\dagger}] = 0. \tag{6.8}$$

It is possible to formulate the uncertainty relation between the photon number and the phase parameter ϕ. For this purpose a and $a^{\dagger}$ are expressed as

$$a = e^{i\phi} N^{1/2} \quad \text{and} \quad a^{\dagger} = N^{1/2} e^{-i\phi}, \tag{6.9}$$

where both $N^{1/2}$ and ϕ are assumed to be Hermitian operators. In terms of these polar variables, the commutation relation $[a, a^{\dagger}] = 1$ of equation (6.8) could be transformed into [3, 9]

$$[N, \phi] = i, \tag{6.10}$$

with the uncertainty relation

$$(\Delta N)(\Delta \phi) \geqslant 1. \tag{6.11}$$

On the other hand, while the above uncertainty relation is consistent with the real world, the commutation relation of equation (6.10) is not consistent with the condition that both ϕ and N be Hermitian [10, 13, 15].

An uncertainty relation without a commutation relation is not new in physics [3, 10]. In addition to Heisenberg's position–momentum uncertainty relations, there is an uncertainty relation between the time and energy variables. However, since the time variable is a c-number, the commutator between the time variable and the Hamiltonian vanishes:

$$[t, H] = 0, \tag{6.12}$$

even though the relation

$$(\Delta t)(\Delta E) \geqslant 1 \tag{6.13}$$

is firmly established and is universally observed. The relation given in equation (6.13) is called the c-number time–energy uncertainty relation [3].

Now, the question we have to address is whether the phase–intensity relation of equation (6.11) is a manifestation of the c-number time–energy uncertainty relation. In order to tackle this issue, let us consider a plane light wave traveling along the z-direction, which is commonly described as

$$Ae^{ik(z-t)}, \tag{6.14}$$

where A is the component of a vector perpendicular to the direction of propagation. Relaxing the localization condition, this describes the light wave corresponding to $|A|^2$ photons with momentum k or energy ω. Since this light is monochromatic, the uncertainty in energy is caused by the photon number $|A|^2$ or N:

$$\Delta E = \omega(\Delta N). \tag{6.15}$$

On the other hand, the uncertainty in phase is that of $k(z - t)$. If the photon energy is constant then

$$\Delta\phi = \omega(\Delta(z - t)). \tag{6.16}$$

If the measurement is made only on the time variable, the uncertainty in $(z - t)$ is the uncertainty in t: $\Delta t = \Delta(z - t)$. Combining equations (6.11), (6.13), and (6.15), it is easily seen that phase–intensity relation [8]

$$(\Delta\phi)(\Delta N) = (\Delta t)(\Delta E), \tag{6.17}$$

is the same as the time–energy uncertainty relation.

It will be more revealing to study this phase-number relation using the Wigner phase-space distribution function, where both the position and momentum variables are c-numbers. We shall discuss this aspect in section 6.5.

6.2 Baker–Campbell–Hausdorff relation

In all branches of physics, we often encounter exponentiation of operators and products of exponentiations. We then quote the mathematical theorem commonly known as the Baker–Campbell–Hausdorff relation. In this section, we derive this relation which is used often in the physics literature and will be used here, including in sections 6.3 and 6.4 of this chapter. The discussion given in this section is based on Miller's book on symmetry groups and their applications [14].

If A is an operator, its exponentiation is defined as

$$e^A = \sum_n \frac{1}{n!}A^n, \tag{6.18}$$

assuming that the series converges. Thus, for two operators A and B, the product of their exponentiations $e^A e^B$ is defined according to the above definition. If A and B commute with each other

$$e^A e^B = e^{(A+B)}. \tag{6.19}$$

If they do not commute, the above relation is not necessarily valid. It is governed by the Baker–Campbell–Hausdorff (BCH) formula.

Let us start with the product

$$e^A B e^{-A}, \tag{6.20}$$

which we encounter very often in physics. In order to calculate this quantity, we consider

$$B(t) = e^{tA} B e^{-tA}. \tag{6.21}$$

Then $B(0) = B$, and $B(t)$ is the quantity to be calculated, where the latter can be expanded as

$$B(t) = \sum_{n=0}^{\infty} \frac{1}{n!} C_n t^n. \tag{6.22}$$

Then, in terms of the infinite series equation above, (6.21) becomes

$$e^A B e^{-A} = \sum_{n=0}^{\infty} \frac{1}{n!} C_n. \tag{6.23}$$

The remaining task is to calculate the coefficients C_n:

$$C_n = \frac{d^n}{dt^n} B(t) \mid_{t=0}. \tag{6.24}$$

The derivative of $B(t)$ with respect to t is

$$\dot{B}(t) = A e^{tA} B e^{-tA} - e^{tA} B e^{-tA} A = [A, B(t)]. \tag{6.25}$$

By repeating this process, it can be shown that

$$C_n = [A, [A, [A, ...[A, B], ...] \ (n \text{ times}). \tag{6.26}$$

For simplicity, the above quantity will be denoted as $(\mathbf{A_d}^n(A)B)$. Consequently, we have

$$e^A B e^{-A} = \exp{[(\mathbf{A_d}(A)B)]}. \tag{6.27}$$

As we shall see in the rest of this chapter, this formula plays the central role in the formalism of coherent and squeezed states.

We also encounter in physics the form

$$e^{C(t)} \frac{d}{dt} e^{-C(t)}, \tag{6.28}$$

where $C(t)$ is an operator or matrix analytic in t. In order to compute this quantity, let us set

$$L(s, t) = e^{sC(t)} \frac{d}{dt} e^{-sC(t)}. \tag{6.29}$$

Then $L(0, t) = 0$, and $L(1, t)$ is the quantity to be calculated. $L(s, t)$ can be expanded as

$$L(s, t) = \sum_{n=0}^{\infty} \frac{1}{n!} D_n(t) s^n. \tag{6.30}$$

Now from equation (6.29),

$$\frac{\partial}{\partial s} L(s, t) = C(t)L(s, t) - L(s, t)C(t) - \dot{C}(t) \tag{6.31}$$

and

$$\left(\frac{\partial}{\partial s}\right)^n L(s, t) = (\mathbf{A_d}^n(C)L(s, t)) - (\mathbf{A_d}^{n-1}(C)\dot{C}(t)). \tag{6.32}$$

Therefore, $D_0(t) = 0$ and

$$D_n(t) = -(\mathbf{A_d}^{n-1}(C)\dot{C}(t)), \tag{6.33}$$

for n greater than zero. This leads to

$$e^{C(t)} \frac{d}{dt} e^{-C(t)} = \sum_{n=0}^{\infty} \frac{1}{(n+1)!} (\mathbf{A_d}^n(C)\dot{C}(t)). \tag{6.34}$$

This series expansion can be written as

$$e^{C(t)} \frac{d}{dt} e^{-C(t)} = -f(\mathbf{A_d}(C))\dot{C}(t), \tag{6.35}$$

where

$$f(z) = \frac{e^z - 1}{z} = 1 + \frac{z}{2!} + \frac{z^2}{3!} + \cdots. \tag{6.36}$$

If we define another function $g(z)$ as

$$g(z) = \frac{\ln(z)}{z - 1} = 1 + \frac{1}{2}(1 - z)^2 + \frac{1}{3}(1 - z)^3 + \cdots, \tag{6.37}$$

then

$$\{f(\ln(z))\}^{-1} = g(z), \tag{6.38}$$

which will be useful in deriving the BCH formula.

We are now ready to prove the theorem of Baker, Campbell, and Hausdorff, which states that, for A, B, and $C = \ln(e^A e^B)$,

$$C = B + \int_0^1 g\{\exp(t[\mathbf{A_d}(A)])\exp[\mathbf{A_d}(B)]\} (A)dt. \tag{6.39}$$

According to this theorem

$$C = A + B + \frac{1}{2}[A, B] + \frac{1}{12}[A, [A, B]] - \frac{1}{12}[B, [B, A]] + \cdots. \tag{6.40}$$

If $[A, B]$ commutes with both A and B, then

$$C = A + B + \frac{1}{2}[A, B],\tag{6.41}$$

which leads to

$$e^A e^B = e^{(A+B)} e^{(1/2)[A, B]}.\tag{6.42}$$

This is the form of the BCH formula seen most often in physics, particularly in the theory of coherent states of light.

Let us prove equation (6.39). Let $C(t) = \ln(e^{tA}e^B)$. Then for an operator H,

$$\begin{aligned}(\exp[\mathbf{A_d}(C(t))])H &= e^{tA}(e^B H e^{-B})e^{-tA}\\ &= (\exp[\mathbf{A_d}(tA)])(\exp[\mathbf{A_d}(B)])(H).\end{aligned}\tag{6.43}$$

Thus

$$\mathbf{A_d}[C(t)] = \ln(\exp[\mathbf{A_d}(tA)]\exp[\mathbf{A_d}(B)]).\tag{6.44}$$

By definition

$$e^{C(t)}\frac{d}{dt}e^{-C(t)} = -e^{tA}e^B e^{-B}e^{-tA}A = -A.\tag{6.45}$$

Hence, according to equation (6.35),

$$f(\mathbf{A_d}(C(t)))\dot{C} = A,\tag{6.46}$$

which, according to equations (6.38) and (6.44), becomes

$$\dot{C}(t) = g(\exp[\mathbf{A_d}(tA)]\exp[\mathbf{A_d}(B)])(A).\tag{6.47}$$

Therefore

$$C(t) = B + \int_0^t g\{\exp(t'[\mathbf{A_d}(A)])\exp[\mathbf{A_d}(B)]\}(A)dt',\tag{6.48}$$

which takes into account $C(0) = B$. When $t = 1$, the above formula leads to the theorem of Baker, Campbell, and Hausdorff given in equation (6.39).

6.3 Coherent states

In this chapter, we are interested in the coherent state defined as [7, 12, 17]

$$|\alpha\rangle = e^{(-|\alpha|^2/2)}\sum_n \frac{\alpha^n}{\sqrt{n!}}|n\rangle,\tag{6.49}$$

where α is a complex number, which may be written in terms of two real numbers α_1 and α_2 as

$$\alpha = \alpha_1 + i\alpha_2.\tag{6.50}$$

This state is normalized

$$\langle\alpha|\alpha\rangle = 1.\tag{6.51}$$

The photon number probability distribution, $P_n(n) = |\langle n|\alpha\rangle|^2$, for the n-photon state is

$$P_n(n) = (\alpha\alpha^*)^n \exp(-\alpha\alpha^*)/n!. \qquad (6.52)$$

This means that the number of photons in the coherent state has a Poisson distribution. Ideal lasers can be formulated in terms of coherent states due to this very aspect of coherent states [1, 11].

The expectation value of the number operator $\langle N \rangle$ is

$$\langle\alpha|N|\alpha\rangle = \sum_n nP_n(n) = \exp(-\alpha\alpha^*)\sum_n \frac{n}{n!}(\alpha\alpha^*)^n, \qquad (6.53)$$

which simplifies as

$$\langle N \rangle = \alpha\alpha^*. \qquad (6.54)$$

At this point it is worth noting that $|\alpha\rangle$ is an eigenstate of a, while it is the left eigenstate of $a^\dagger$, namely

$$\begin{aligned}a|\alpha\rangle &= \alpha|\alpha\rangle, \\ \langle\alpha|a^\dagger &= \alpha^*\langle\alpha|.\end{aligned} \qquad (6.55)$$

In view of the above properties of the coherent state and equation (6.4) the expectation value of the number operator can be found more simply

$$\langle N \rangle = \langle\alpha|a^\dagger a|\alpha\rangle = \alpha^*\alpha\langle\alpha|\alpha\rangle = \alpha^*\alpha. \qquad (6.56)$$

The expectation value of N^2 is also useful in quantum optics. The eigenvalue of this operator on $|n\rangle$ is of course n^2, which can be written as $n^2 = n(n-1) + n$. This leads to

$$\langle N^2 \rangle = (\alpha\alpha^*)^2 + \alpha\alpha^*. \qquad (6.57)$$

From this, we can calculate the variance of the number operator defined as

$$(\Delta N)^2 = \langle N^2 \rangle - \langle N \rangle^2 = \alpha\alpha^*. \qquad (6.58)$$

In terms of the position variable x, the n-photon state can be represented by a harmonic oscillator wave function. We shall use the notation

$$\langle x|n\rangle = \left(\frac{1}{\sqrt{\pi}\,2^n n!}\right)^{1/2} H_n(x)\exp(-x^2/2). \qquad (6.59)$$

This n-photon state can also be written in terms of the momentum wave function, which takes the form

$$\langle p|n\rangle = (-i)^n \left(\frac{1}{\sqrt{\pi}\,2^n n!}\right)^{1/2} H_n(p)\exp(-p^2/2). \qquad (6.60)$$

The coherent state $|\alpha\rangle$ is not an eigenstate of the number operator. In terms of the Hermite polynomials, it can be written as

$$\langle x|\alpha\rangle = \left(\frac{1}{\pi}\right)^{1/4} e^{-\alpha\alpha^*/2} \exp\left(\frac{-x^2}{2}\right) \sum_n \frac{\alpha^n}{n!}\left(\frac{1}{2}\right)^{n/2} H_n(x), \tag{6.61}$$

while, in terms of the momentum variable p, it becomes

$$\langle p|\alpha\rangle = \left(\frac{1}{\pi}\right)^{1/4} e^{-\alpha\alpha^*/2} \exp\left(\frac{-p^2}{2}\right) \sum_n \frac{(-i\alpha)^n}{n!}\left(\frac{1}{2}\right)^{n/2} H_n(p). \tag{6.62}$$

We are now interested in writing these functions as Gaussian functions. For this purpose, let us recall that the coherent state $|\alpha\rangle$ is an eigenstate of the annihilation operator and refer to equation (6.55). Therefore, it should satisfy the differential equation

$$\frac{1}{\sqrt{2}}\left(x + \frac{\partial}{\partial x}\right)\langle x|\alpha\rangle = \alpha\langle x|\alpha\rangle. \tag{6.63}$$

Its normalized solution is

$$\langle x|\alpha\rangle = \left(\frac{1}{\pi}\right)^{1/4} \exp\left\{-\frac{1}{2}\left[(x - \sqrt{2}\alpha)^2 + 2\alpha_2^2\right]\right\}, \tag{6.64}$$

where α_2 is the imaginary component of the complex number α, as given in equation (6.50). If we Fourier-transform this expression, it becomes the momentum wave function:

$$\langle p|\alpha\rangle = \left(\frac{1}{\pi}\right)^{1/4} \exp\left\{-\frac{1}{2}\left[(p + i\sqrt{2}\alpha)^2 + 2\alpha_1^2\right]\right\}. \tag{6.65}$$

When α is a real number, it is a displaced ground-state harmonic oscillator wave function, attaining its maximum value at $x = \sqrt{2}\alpha$. When α is complex, it is a ground-state wave function whose origin is displaced to a complex value. However, the probability distribution becomes

$$|\langle x|\alpha\rangle|^2 = \left(\frac{1}{\pi}\right)^{1/2} \exp\left\{-\left(x - \sqrt{2}\alpha_1\right)^2\right\}. \tag{6.66}$$

For the momentum wave function

$$|\langle p|\alpha\rangle|^2 = \left(\frac{1}{\pi}\right)^{1/2} \exp\left\{-\left(p - \sqrt{2}\alpha_2\right)^2\right\}. \tag{6.67}$$

Indeed, the probability distribution is a Gaussian function for all possible values of α. This leads to

$$\langle(\Delta x)^2\rangle\langle(\Delta p)^2\rangle = 1/4. \tag{6.68}$$

It is now clear that the coherent state is a minimum-uncertainty state.

Next, let us consider the orthogonality relations of the coherent states. The eigenvalue α has two real components. Thus the sum over this continuous eigenvalue is necessarily an integral over the two-dimensional space of α_1 and α_2. With this point in mind, let us look more closely at the integration

$$\frac{1}{\pi} \int |\alpha\rangle\langle\alpha| \, d\alpha_1 \, d\alpha_2. \tag{6.69}$$

The integral over the two-dimensional space can be converted into a polar coordinate with

$$r = (\alpha_1^2 + \alpha_2^2)^{1/2} \text{ and } \phi = \tan^{-1}(\alpha_2/\alpha_1). \tag{6.70}$$

Then

$$\alpha\alpha^* = r^2, \qquad \alpha^n = r^n e^{in\phi}, \quad \text{and} \quad d\alpha_1 d\alpha_2 = rdrd\phi. \tag{6.71}$$

As a result

$$\frac{1}{\pi} \int |\alpha\rangle\langle\alpha| d\alpha_1 d\alpha_2 = \sum_n \left\{ \frac{2}{n!} \int_0^\infty r^{(2n+1)} e^{-r^2} dr \right\} |n\rangle\langle n|$$
$$= \sum_n |n\rangle\langle n|, \tag{6.72}$$

with

$$\frac{2}{n!} \int_0^\infty r^{(2n+1)} e^{-r^2} dr = 1. \tag{6.73}$$

Thus, the integral of equation (6.69) is an identity operator.

Although the number states are orthonormal, as seen from equation (6.1), it should not be taken for granted that this property holds for the coherent states. For this purpose consider $\langle\beta|\alpha\rangle$. Taking into account equation (6.49), we have

$$\langle\beta|\alpha\rangle = \exp\left(-(\alpha\alpha^* + \beta\beta^*)/2\right) \sum_n \frac{1}{n!} (\alpha\beta^*)^n$$
$$= \exp\left(-(\alpha\alpha^* + \beta\beta^* - 2\alpha\beta^*)/2\right), \tag{6.74}$$

which leads to

$$|\langle\beta|\alpha\rangle|^2 = \exp\left(-|\alpha - \beta|^2\right). \tag{6.75}$$

This means that $|\alpha\rangle$ and $|\beta\rangle$ have overlaps and, hence, the coherent state is not a complete orthonormal state, but over-complete.

6.4 Symmetry of coherent states

Let us introduce an operator of the form

$$U(\alpha) = \exp\left(\alpha a^\dagger - \alpha^* a\right). \tag{6.76}$$

This operator is unitary $[U(\alpha)]^\dagger = [U(\alpha)]^{-1} = U(-\alpha)$, as $e^A e^{-A} = I$, for any operator A. Making use of the special case of the BCH formula for two operators whose commutation is a c-number, as given in equation (6.42), $U(\alpha)$ becomes

$$U(\alpha) = e^{-\alpha\alpha^*/2} e^{\alpha a^\dagger} e^{-\alpha^* a}. \tag{6.77}$$

An operator in the form as in equation (6.76) is usually called the displacement operator, since it has the capacity to displace a localized state in the phase space by the magnitude of α.

Now, consider the action of $U(\alpha)$ on the vacuum state $|0\rangle$:

$$U(\alpha)|0\rangle = e^{-\alpha\alpha^*/2} e^{\alpha a^\dagger} e^{-\alpha^* a} |0\rangle. \tag{6.78}$$

Since $\exp(-\alpha^* a)|0\rangle = |0\rangle$, this expression becomes

$$U(\alpha)|0\rangle = e^{-\alpha\alpha^*/2} e^{\alpha a^\dagger} |0\rangle$$
$$= e^{-\alpha\alpha^*/2} \sum_n \frac{\alpha^n}{n!} (a^\dagger)^n |0\rangle. \tag{6.79}$$

Then, in view of equation (6.6), it simplifies as

$$U(\alpha)|0\rangle = e^{-\alpha\alpha^*/2} \sum_n \frac{\alpha^n}{\sqrt{n!}} |n\rangle, \tag{6.80}$$

where the right-hand side is as in equation (6.49). Therefore, we have the coherent state as a unitary transformation acting on the vacuum

$$U(\alpha)|0\rangle = |\alpha\rangle. \tag{6.81}$$

An alternative expression for the coherent state is commonly found in the literature [12], where it takes the form

$$|\alpha\rangle = e^{(-|\alpha|^2/2)} \sum_n \frac{\alpha^n}{n!} (a^\dagger)^n |0\rangle, \tag{6.82}$$

which can easily be deduced from equation (6.79).

Next, let us see the effect of the number operator $N = a^\dagger a$. This operator is Hermitian, and its exponentiation

$$M(\phi) = e^{-i\phi N} \tag{6.83}$$

performs a unitary transformation. Its application to the coherent state leads to

$$M(\phi)|\alpha\rangle = e^{-\alpha\alpha^*/2} \sum_n \frac{(e^{-i\phi}\alpha)^n}{n!} (a^\dagger)^n |0\rangle = |e^{-i\phi}\alpha\rangle. \tag{6.84}$$

Thus the operation of $M(\phi)$ decreases the phase angle of α by ϕ, and the operator N generates the rotation of the phase angle of α. The commutation relations of N with a and $a^\dagger$ are

$$[a, N] = a, \qquad [a^\dagger, N] = -a^\dagger, \qquad \text{and} \qquad [a, a^\dagger] = 1. \tag{6.85}$$

The group generated by N, a, $a^\dagger$, and 1 is called the harmonic oscillator group. Hence, the coherent state is a representation of the harmonic oscillator group.

Since the creation and annihilation operators are not Hermitian, we are led to consider

$$x = \frac{1}{\sqrt{2}}(a^\dagger + a) \quad \text{and} \quad p = \frac{-i}{\sqrt{2}}(a^\dagger - a), \tag{6.86}$$

that are Hermitian. They satisfy the familiar Heisenberg commutation relation

$$[x, p] = i. \tag{6.87}$$

In terms of these Hermitian operators, the number operator becomes

$$N = \frac{1}{2}(p^2 + x^2 - 1), \tag{6.88}$$

which is also Hermitian. With x and p, its commutation relations are

$$[N, x] = -ip \qquad \text{and} \qquad [N, p] = ix. \tag{6.89}$$

The operator $U(\alpha)$ can then be written as

$$U(\alpha) = \exp\left\{ i\left(\sqrt{2}\,\alpha_2\right)x - i\left(\sqrt{2}\,\alpha_1\right)p \right\}, \tag{6.90}$$

where α_1 and α_2 are two real independent parameters. Using the BCH formula, we can transform this into

$$U(\alpha) = \exp\left(i\alpha_1\alpha_2\right) \exp\left\{ i\left(\sqrt{2}\,\alpha_2\right)x \right\} \exp\left\{ -i\left(\sqrt{2}\,\alpha_1\right)\frac{\partial}{\partial x} \right\}. \tag{6.91}$$

If this is applied to the vacuum state $|0\rangle$, the wave function becomes

$$\langle x | U(\alpha) | 0 \rangle = \left(\frac{1}{\pi}\right)^{1/4} \exp\left\{ i\alpha_1\alpha_2 + i\left(\sqrt{2}\,\alpha_2\right)x \right\} \exp\left\{ -\left(x - \sqrt{2}\,\alpha_1\right)^2/2 \right\}, \tag{6.92}$$

being in the form of the expression in equation (6.64) for $\langle x|\alpha\rangle$. It is thus clear that p generates translations of the wave function along the real x-axis, and $-x$ generates the addition of the imaginary number to α. Thus p and x appear to generate the translation group in the two-dimensional plane of α_1 and α_2. However, since they do not commute with each other, they cannot be regarded as the translation generators.

In order to clarify this problem, let us consider two repeated applications of U. Again, from the BHC relation given in section 6.2,

$$U(\beta)U(\alpha) = e^{i(Im(\beta\alpha^*))}U(\alpha + \beta). \tag{6.93}$$

The complex parameter α is additive in the sense that the right-hand side is $U(\alpha + \beta)$ multiplied by a factor of unit modulus. The unitarity is preserved, consequently

$$U(\beta)U(\alpha)U(-\beta) = e^{2i(\text{Im}(\beta\alpha^*))}U(\alpha). \tag{6.94}$$

Let us see whether this relation can be derived from $[U(\beta)\,a\,U(-\beta)]$ and $[U(\beta)\,a^\dagger\,U(-\beta)]$. First, we write $[U(\beta)\,a\,U(-\beta)]$ as

$$U(\beta)\,a\,U(-\beta) = e^{(\beta a^\dagger - \beta^* a)}a e^{-(\beta a^\dagger - \beta^* a)}. \tag{6.95}$$

Let us then go back to equation (6.27). Since

$$[a^\dagger, a] = -1 \quad \text{and} \quad [a^\dagger, [a^\dagger, a]] = 0, \tag{6.96}$$

we have

$$U(\beta)\,a\,U(-\beta) = a - \beta. \tag{6.97}$$

Likewise

$$U(\beta)\,a^\dagger\,U(-\beta) = \hat{a}^\dagger - \beta^*. \tag{6.98}$$

Therefore

$$U(\beta)U(\alpha)U(-\beta) = \left\{ e^{-(\alpha\beta^* - \alpha^*\beta)} \right\} U(\alpha). \tag{6.99}$$

This result is consistent with equation (6.93).

The story is similar for $M(\phi)\,a\,M(-\phi)$ and $M(\phi)\,a^\dagger\,M(-\phi)$. Again, let us go back to equation (6.27), and note

$$[N, a] = -a \quad \text{and} \quad [N, [N, a]] = (-1)^2 a, \tag{6.100}$$

which leads to

$$[N, [N, \cdots[N, [N, a]\cdots] \quad \text{(n times)} = (-1)^n a. \tag{6.101}$$

Therefore,

$$M(\phi)\,a\,M(-\phi) = e^{-i\phi N}\,a\,e^{i\phi N} = e^{i\phi}a. \tag{6.102}$$

Likewise

$$M(\phi)\,a^\dagger\,M(-\phi) = e^{-i\phi}a^\dagger. \tag{6.103}$$

The symmetry of the coherent state is dictated by the transformations $U(\alpha)$ and $M(\phi)$. $M(\phi)$ is clearly a rotation operator in the complex plane of α. $U(\beta)$ appears as a translation operator in equations (6.97) and (6.98). However, equation (6.99) shows that it does not represent a commutative group. Thus $U(\alpha)$ represents a multiplier representation of the translation group. This point will be examined further in section 6.5 where we shall study the symmetry problems using the Wigner function.

6.5 Coherent states in phase space

We have seen in sections 6.3 and 6.4 that the wave function corresponding to the coherent state is basically Gaussian, and that its center of distribution depends on the parameter α. For the distribution in the x variable, its center depends on the real part of the α variable, and for the distribution in the p variable its center depends on the imaginary part of α.

This leads us to study the coherent state using the Wigner phase-space picture. The Wigner function is defined as [18]

$$W(x, p) = \frac{1}{\pi} \int e^{2i(p \cdot x')} \langle x - x'|\alpha \rangle^* \langle x + x'|\alpha \rangle \, dx'. \tag{6.104}$$

This is a function defined in the two-dimensional space of x and p.

The evaluation of this integral leads to the Wigner function of the form [11]

$$W(\alpha; x, p) = \frac{1}{\pi} \exp\left\{-(x - a)^2 - (p - b)^2\right\}, \tag{6.105}$$

where

$$a = \sqrt{2}\, Re(\alpha) \quad \text{and} \quad b = \sqrt{2}\, Im(\alpha). \tag{6.106}$$

This Wigner function reproduces the probability distribution functions of equation (6.66). If integrated out in the p variable the distribution is in the x variable:

$$|\langle x|\alpha \rangle|^2 = \int W(\alpha; x, p) \, dp, \tag{6.107}$$

while the probability distribution in the p variable is

$$|\langle p|\alpha \rangle|^2 = \int W(\alpha; x, p) \, dx. \tag{6.108}$$

The Wigner function of equation (6.105) leads to

$$|\langle \beta|\alpha \rangle|^2 = \int W(\beta; x, p) W(\alpha; x, p) \, dx \, dp$$
$$= \exp\left(-|\alpha - \beta|^2\right), \tag{6.109}$$

which is consistent with one of the basic properties of the coherent state, as given in equation (6.75).

If $\alpha = 0$, both a and b vanish, the Wigner function becomes

$$W(0; x, p) = \frac{1}{\pi} \exp\left\{-(x^2 + p^2)\right\}, \tag{6.110}$$

which is concentrated within a circular region around the origin. This is the Wigner function for the vacuum state.

We can obtain the Wigner function of equation (6.105) by making canonical transformations of the vacuum-state Wigner function. More specifically, the Wigner function of equation (6.105) is a translation of the vacuum state along the x-axis by a and along the p-axis by b.

In section 6.3 we obtained the coherent state by applying a unitary transformation to the vacuum state. The generator transformation of this unitary transformation was $i(\alpha a^\dagger - \alpha^* a)$, with one complex parameter α having two real numbers.

In the Wigner phase space, we are now working with translations in the two orthogonal directions, namely x and p. These generators can take the form

$$P_1 = -i\frac{\partial}{\partial x} \quad \text{and} \quad P_2 = -i\frac{\partial}{\partial p}, \tag{6.111}$$

in this phase-space picture in which P_1 and P_2 commute with each other.

In addition, we can consider the rotation generator defined as

$$L = -i\left\{x\frac{\partial}{\partial p} - p\frac{\partial}{\partial x}\right\}. \tag{6.112}$$

The Wigner function given in equation (6.105) is localized within the circle:

$$(x - a)^2 + (p - b)^2 = 1. \tag{6.113}$$

Thus, all the instruments developed in chapter 5 to deal with canonical transformations of a circle are applicable to the Wigner function for the coherent state.

It was noted in chapter 3 that P_1, P_2, and L are the generators of the two-dimensional Euclidean group. Rotations around the origin form the one-parameter subgroup generated by L. Translations generated by P_1 and P_2 form a two-parameter subgroup. The translation operator applicable to the Wigner function is

$$T(\alpha) = \exp\left(-a\frac{\partial}{\partial x} - b\frac{\partial}{\partial p}\right), \tag{6.114}$$

which leads to

$$T(\alpha)W(0; q, p) = W(\alpha; q, p). \tag{6.115}$$

If we multiply α by $e^{i\phi}$, the circle corresponding to equation (6.113) becomes rotated around the origin, and the resulting equation is

$$(x - a')^2 + (p - b')^2 = 1, \tag{6.116}$$

where

$$\begin{pmatrix} a' \\ b' \end{pmatrix} = \begin{pmatrix} \cos\phi & -\sin\phi \\ \sin\phi & \cos\phi \end{pmatrix} \begin{pmatrix} a \\ b \end{pmatrix}. \tag{6.117}$$

The rotation operator applicable to the Wigner function is

$$R(\phi) = \exp\left\{-\phi\left(x\frac{\partial}{\partial p} - p\frac{\partial}{\partial x}\right)\right\}, \tag{6.118}$$

which leads to

$$R(\phi)W(\alpha; x, p) = W(\alpha e^{i\phi}; x, p). \tag{6.119}$$

As is shown in figure 6.1, the result of the above rotation is only a translation of the circle centered at the origin to (a', b'). Therefore, do we really need the rotation operator? In order to answer this question, let us write the transformation operators as three-by-three matrices applicable to the three-dimensional space of $(x, p, 1)$. They can be written as

$$T(\alpha) = \begin{pmatrix} 1 & 0 & a \\ 0 & 1 & b \\ 0 & 0 & 1 \end{pmatrix} \quad \text{and} \quad R(\phi) = \begin{pmatrix} \cos\phi & -\sin\phi & 0 \\ \sin\phi & \cos\phi & 0 \\ 0 & 0 & 1 \end{pmatrix}. \tag{6.120}$$

Then these matrices lead to

$$T(\beta)T(\alpha) = T(\alpha + \beta) \quad \text{and} \quad R(\phi)T(\alpha) = T(\alpha')R(\phi). \tag{6.121}$$

Thus, the most general form of the transformation matrix can be written as $T(\alpha'')R(\phi)$. However, the rotation on the circle centered at the origin does not change the circle. The net effect is the $T(\alpha'')$, as illustrated in figure 6.1.

In this section, we have demonstrated that the symmetry of the coherent state is that of the two-dimensional Euclidean group in the Wigner phase-space representation. Let us go back to section 6.4 where we discussed the same symmetry in the complex α-plane. The symmetry group is not Euclidean there, because the x and p operations do not commute with each other. It is interesting to note that these two groups give the same result.

6.6 Single-mode squeezed states

In addition to the rotation matrix of equation (6.120), we can consider

$$B(\eta) = \begin{pmatrix} e^\eta & 0 & 0 \\ 0 & e^{-\eta} & 0 \\ 0 & 0 & 1 \end{pmatrix}. \tag{6.122}$$

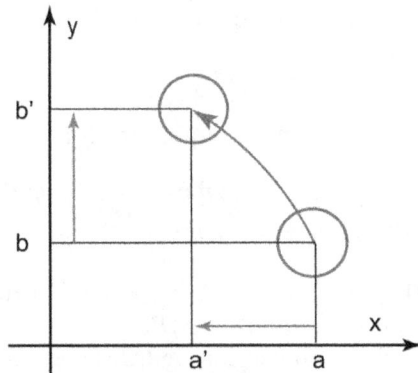

Figure 6.1. Euclidean transformations in phase space. A rotation around the origin results in a translation [11].

This matrix performs a squeeze transformation:

$$x \to e^{\eta}x \quad \text{and} \quad p \to e^{-\eta}p. \tag{6.123}$$

Then, as in the case of the rotation matrix, we have

$$B(\eta)T(\alpha) = T(\alpha')B(\eta), \tag{6.124}$$

with

$$a' = e^{\eta}a \quad \text{and} \quad b' = e^{-\eta}b. \tag{6.125}$$

Let us next change the direction of the squeeze. This can be achieved from the rotation of the $B(\eta)$ matrix

$$S(\phi, \eta) = R(\phi)B(\eta)R^{-1}(\phi). \tag{6.126}$$

This transformation matrix takes the form

$$S(\phi, \eta) = \begin{pmatrix} \cosh\eta + (\sinh\eta)\cos(2\phi) & (\sinh\eta)\sin(2\phi) & 0 \\ (\sinh\eta)\sin(2\phi) & \cosh\eta - (\sinh\eta)\cos(2\phi) & 0 \\ 0 & 0 & 1 \end{pmatrix}. \tag{6.127}$$

Indeed, this combination of squeeze and rotation is generated by the two squeeze generators

$$S_1 = -i\left(x\frac{\partial}{\partial x} - p\frac{\partial}{\partial p}\right) \quad \text{and} \quad S_2 = -i\left(x\frac{\partial}{\partial p} + p\frac{\partial}{\partial x}\right), \tag{6.128}$$

which together with the rotation generator L of equation (6.112), form the following set of commutation relations:

$$[L, S_1] = i\,S_2, \quad [L, S_2] = i\,S_1, \quad \text{and} \quad [S_1, S_2] = i\,L. \tag{6.129}$$

This is the Lie algebra of the $O(2,1)$ group which was discussed in chapter 3.

From the relation $R(\phi)T(\alpha) = T(\alpha')R(\phi)$ for the rotation matrix of equation (6.119) and $B(\eta)T(\beta) = T(\beta')B(\eta)$ of equation (6.124), it is possible to write

$$S(\phi, \eta)T(\alpha) = T(\alpha'')S(\phi, \eta), \tag{6.130}$$

where a'' and b'' for α'' are, according to equation (6.127),

$$a'' = [\cosh\eta + (\sinh\eta)\cos(2\phi)]\,a + (\sinh\eta)\sin(2\phi)\,b,$$
$$b'' = (\sinh\eta)\sin(2\phi)\,a + [\cosh\eta - (\sinh\eta)\cos(2\phi)]\,b. \tag{6.131}$$

If $S(\phi, \eta)$ is applied to the vacuum state, or the circle centered at the origin, the rotation gives no effect, and it becomes $R(\phi)B(\eta)$, namely a squeeze followed by a rotation. This squeezed circle or ellipse can be translated along the radial direction, as illustrated in figure 6.2.

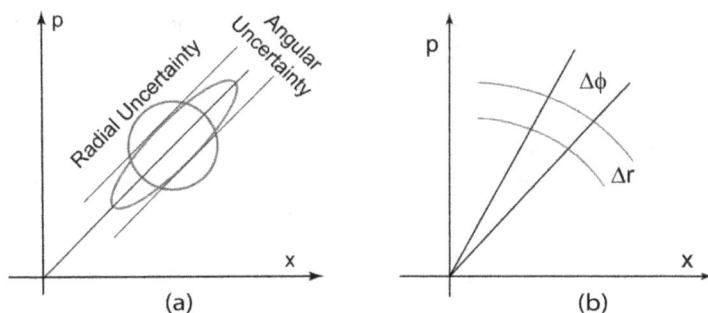

Figure 6.2. Radial squeeze in the phase space: (a) shows how the coherent state can be radially squeezed to produce the variations of radius and angle corresponding to the uncertainty relation between the number and phase illustrated in (b). The number–phase uncertainty relation was discussed in section 6.1.

The phase-number uncertainty relation $(\Delta\phi)(\Delta N) \cong 1$ can be canonically transformed into the phase space of x and p. Let us consider the transformation from the Cartesian coordinate system into a polar coordinate system with [5]

$$N = \frac{1}{2}(x^2 + p^2 - 1) \quad \text{and} \quad \phi = \tan^{-1}(p/x). \qquad (6.132)$$

This is a canonical transformation from the Cartesian coordinates of x and p with

$$\frac{\partial N}{\partial x}\frac{\partial \phi}{\partial p} - \frac{\partial \phi}{\partial x}\frac{\partial N}{\partial p} = 1. \qquad (6.133)$$

The c-number uncertainty relation in x and p in the phase-space picture of quantum mechanics is canonically equivalent to the uncertainty in ϕ and N. The uncertainty relation $(\Delta x)(\Delta p) = 1$ is translated into $(\Delta\phi)(\Delta N) = 1$.

References

[1] Arecchi F T 1965 Measurement of the statistical distribution of Gaussian and laser sources *Phys. Rev. Lett.* **15** 912–6

[2] Barnett S M and Radmore P M 2005 *Methods in Theoretical Quantum Optics* (*Oxford Series in Optical and Imaging Sciences*) vol 15 (Oxford: Clarendon)

[3] Dirac P A M 1927 The quantum theory of the emission and absorption of radiation *Proc. R. Soc.* A **114** 243–65

[4] Dirac P A M 1943 Quantum electrodynamics *Commun. Dublin Inst. Adv. Stud.* A **1** 36

[5] Dirl R, Kasperkovitz P, and Moshinsky M 1988 Wigner distribution functions and the representation of a non-bijective canonical transformation in quantum mechanics *J. Phys. A: Math. Gen.* **21** 1835–46

[6] Fock V 1932 Konfigurationsraum und zweite Quantelung *Z. Phys.* **75** 622–47

[7] Glauber R J 1963 Coherent and incoherent states of the radiation field *Phys. Rev.* **131** 2766–88

[8] Goldin E 1982 *Waves and Photons: An Introduction to Quantum Optics* (*Wiley Series in Pure and Applied Optics*) (New York: Wiley)

[9] Heitler W 1984 *The Quantum Theory of Radiation* 3rd edn (New York: Dover)

[10] Hussar P E, Kim Y S, and Noz M E 1985 Time–energy uncertainty relation and Lorentz covariance *Am. J. Phys.* **53** 142–7

[11] Kim Y S and Noz M E 1991 *Phase Space Picture of Quantum Mechanics: Group Theoretical Approach* (*Lecture Notes in Physics* vol 40) (Singapore: World Scientific)

[12] Klauder J R and Skagerstam B S 1985 *Coherent States* (Singapore: World Scientific)

[13] Louisell W H 1973 *Quantum Statistical Properties of Radiation* (New York: Wiley)

[14] Miller W 1972 *Symmetry Groups and Their Applications. Pure and Applied Mathematics* (New York: Academic)

[15] Pegg D T and Barnett S M 1997 Quantum optical phase *J. Mod. Opt.* **44** 225–64

[16] Schrödinger E 1926 Der stetige Übergang von der Mikro- zur Makromechanik *Naturwissenschaften* **14** 664–6

[17] Sudarshan E C G 1963 Equivalence of semiclassical and quantum mechanical descriptions of statistical light beams *Phys. Rev. Lett.* **10** 277–9

[18] Wigner E P 1932 On the quantum correction for thermodynamic equilibrium *Phys. Rev.* **40** 749–59

IOP Publishing

Mathematical Devices for Optical Sciences

Sibel Başkal, Young S Kim, and Marilyn E Noz

Chapter 7

Squeezed states and their symmetries

The two-mode coherent photon state has been the central issue in optical sciences since 1976 [21]. This state is commonly called the squeezed state of light, and is covered in many books on quantum optics [17, 19]. We intend here to point out that the squeezed state has a much broader mathematical basis. The subject is largely based on the group of Lorentz transformations.

Einstein used the Lorentz group to formulate his special theory of relativity, which is the basic language for high-energy particle physics. This role of the Lorentz group was illustrated in figure 2.1. The two-dimensional harmonic oscillator also serves as the basic language for the squeezed state. In 1963 [5], Paul A M Dirac carefully studied the symmetries generated by two oscillators and constructed the symmetry group consisting of ten generators. The Lie algebra of these ten generators is the same as that of the $O(3, 2)$ group, namely the Lorentz group applicable to three space-like and two time-like dimensions. The four-by-four representation of this group is the $Sp(4)$ group, which is the group of canonical transformations of two phase spaces in classical and quantum mechanics [1, 7].

In 1986 [22], Yurke *et al* considered two-mode interferometers satisfying $U(2)$ and $U(1, 1)$ symmetries. Indeed, the symmetry group containing both of these two symmetries is exactly the $O(3, 2)$-like group formulated in 1963 [5].

Earlier, Paul A M Dirac constructed four traceless matrices, namely the γ_0, γ_1, γ_2, and γ_3 matrices for his Dirac equation. With these matrices, it is possible to construct fifteen traceless Dirac matrices. In the Majorana representation [15], all the elements in these matrices are real, leading to four-by-four transformation matrices with real elements.

This means that there are five Dirac matrices which do not lead to canonical transformations. Indeed, the system of fifteen Dirac matrices leads to the Lie algebra for the $O(3, 3)$ Lorentz group.

In section 7.1 we formulate the two-mode state in terms of two harmonic oscillators, and study the transformation properties using two oscillator wave

7-1

functions in section 7.2. In section 7.3 we study the symmetry properties of the two-mode state in terms of the $O(3, 2)$-like Lorentz group introduced by Dirac in 1963 [5]. This group is like the $Sp(4)$ group governing linear canonical transformations for two particles, each with a two-dimensional phase space. These are generated by ten four-by-four matrices with imaginary elements.

On the other hand, there are fifteen four-by-four Dirac matrices. Thus, five of those four-by-four matrices should generate non-canonical transformations. Thus, as is shown in section 7.4, those fifteen four-by-four matrices constitute the Lie algebra for the group $O(3, 3)$, namely the group of Lorentz transformations applicable to three space-like and three time-like directions. However, it is still the group of four-by-four matrices.

In section 7.5 we study this four-dimensional symmetry using Wigner functions applicable to the four-dimensional phase space. In section 7.6 we study the symmetry problems where we apply them to those of two coupled oscillators.

7.1 Two-mode states

The states of two different photons play a pivotal role in modern physics. They are called two-photon coherent states, squeezed states, or two-photon entangled states. Let us write the two-photon state as

$$|n_1, n_2\rangle. \tag{7.1}$$

The most general form for the two-mode state could take the form

$$\sum_{n_1, n_{n_2}} A_{n_1, n_2} |n_1\rangle |n_2\rangle. \tag{7.2}$$

However, the most useful form for the current trend in physics is the case with

$$n_1 = n_2 \quad \text{and} \quad A_{n, n} = \frac{\beta^n}{\sqrt{1 - \beta^2}}. \tag{7.3}$$

Thus, this two-mode state becomes

$$|\beta\rangle = \frac{1}{\sqrt{1 - \beta^2}} \sum_n \beta^n |n, n\rangle, \tag{7.4}$$

where we write $|n\rangle |n\rangle$ as $|n, n\rangle$. Since $|n\rangle = \sqrt{n!} \, [a^\dagger]^n |0\rangle$, this two-mode state takes the form

$$|\beta\rangle = \frac{1}{\sqrt{1 - \beta^2}} e^{\beta a_1^\dagger a_2^\dagger} |0, 0\rangle. \tag{7.5}$$

The question is whether this expression can be written as a unitary transformation. It is of course possible to write this equation as

$$|\beta\rangle = \frac{1}{\sqrt{1-\beta^2}} e^{\beta a_1^\dagger a_2^\dagger} e^{-\beta a_1 a_2} |0, 0\rangle, \tag{7.6}$$

but this is not a unitary transformation. The unitary transformation should take the form

$$|\beta\rangle = e^{\beta[a_1^\dagger a_2^\dagger - a_1 a_2]} |0, 0\rangle, \tag{7.7}$$

and this unitary operation should lead to the two-mode coherent state of equation (7.4). This question will be discussed in detail in section 7.2. Here, a, $a^\dagger$ are the annihilation and creation operators, respectively, as defined in equation (6.3).

The total energy of the state with two photons with different frequencies is clearly

$$\omega_1 n_1 + \omega_2 n_2, \tag{7.8}$$

where ω_1 and ω_2 are the frequencies of the first and second kinds, respectively. Then the energy of the two-mode state of equation (7.4) is

$$(\omega_1 + \omega_2)(1 - \beta^2) \sum_k k\beta^{2k} = (1 - \beta^2)(\omega_1 + \omega_2)\frac{\beta}{2}\left(\frac{\partial}{\partial \beta} \sum_k \beta^{2k}\right), \tag{7.9}$$

which becomes

$$\frac{(\omega_1 + \omega_2)\beta^2}{(1 - \beta^2)} = (\omega_1 + \omega_2)(\sinh \eta)^2. \tag{7.10}$$

The energy is zero for the vacuum state with $\eta = 0$, but it increases as the system becomes squeezed with increasing values of η.

7.2 Unitary transformations

The question here is whether it is possible to obtain the two-mode coherent state by making a unitary transformation of the vacuum state. We shall prove in this section that the unitary transformation of equation (7.7) leads to the series given in equation (7.4). In order to achieve this, we go back to the language of harmonic oscillators, carry out the calculations, and come back to the language of a_i and $a_i^\dagger$.

In terms of the harmonic oscillator wave functions, the series of equation (7.4) can be written as

$$|\beta\rangle = \sqrt{1-\beta^2} \sum_k \chi_k(x_1)\chi_k(x_2), \tag{7.11}$$

and the exponential form in equation (7.7) takes the form

$$\exp\left\{\beta\left(x_1\frac{\partial}{\partial x_2} - x_2\frac{\partial}{\partial x_1}\right)\right\}, \tag{7.12}$$

where a_i and $a_i^\dagger$ in equation (7.7) have been replaced with

$$a_i = \frac{1}{\sqrt{2}}\left(x_i + \frac{\partial}{\partial x_i}\right) \quad \text{and} \quad a_i^\dagger = \frac{1}{\sqrt{2}}\left(x_i - \frac{\partial}{\partial x_i}\right), \tag{7.13}$$

respectively, as defined in equation (1.130).

If this operator is applied to the ground-state wave function, as shown in figure 7.1, the result is

$$|\beta\rangle = \frac{1}{\sqrt{\pi}} \exp\left\{-\frac{1}{2}(y_1^2 + y_2^2)\right\}, \tag{7.14}$$

where

$$\begin{pmatrix} y_1 \\ y_2 \end{pmatrix} = \begin{pmatrix} \cosh\eta & \sinh\eta \\ \sinh\eta & \cosh\eta \end{pmatrix}\begin{pmatrix} x_1 \\ x_2 \end{pmatrix}, \tag{7.15}$$

with

$$\beta = \tanh\eta. \tag{7.16}$$

In terms of the x_1 and x_2 variables, the wave function of equation (7.14) can be written as

$$|\beta\rangle = \frac{1}{\sqrt{\pi}} \exp\left\{-\frac{1}{4}\left[e^{-2\eta}(x_1 + x_2)^2 + e^{2\eta}(x_1 - x_2)^2\right]\right\}. \tag{7.17}$$

The problem is then to write the exponential expression of equation (7.17) as the series

$$|\beta\rangle = \sum_{k,k'} A_{k,k'}\chi_k(x_1)\chi_{k'}(x_2), \tag{7.18}$$

and evaluate the coefficient $A_{k,k'}$, with

$$\sum_{k,k'} |A_{k,k'}|^2 = 1. \tag{7.19}$$

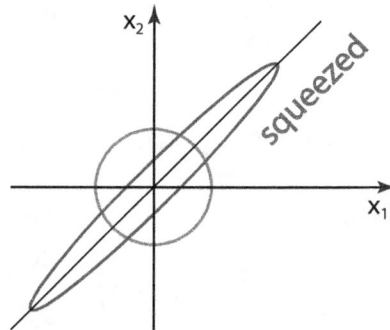

Figure 7.1. Squeezed Gaussian wave function. The ground-state wave function has a circular distribution, and it is appropriate to call the elliptic distribution a *squeezed state* [2].

Let us first write the wave function $\chi_k(x_1)\,\chi_{k'}(x_2)$ which satisfies the differential equation:

$$\frac{1}{2}\left\{\left(x_1^2 - \frac{\partial^2}{\partial x_1^2}\right) - \left(x_2^2 - \frac{\partial^2}{\partial x_2^2}\right)\right\}\chi_k(x_1)\chi_{k'}(x_2) = (k' - k)\chi_k(x_1)\chi_{k'}(x_2). \qquad (7.20)$$

This equation is invariant under the transformation given in equation (7.15), is Lorentz-invariant, and $k - k' = 0$. Hence the series takes the form

$$|\beta\rangle = \sum_k A_k \chi_k(x_1)\chi_k(x_2). \qquad (7.21)$$

This is a sum over a single variable k, with

$$\sum_k |A_k|^2 = 1. \qquad (7.22)$$

This coefficient is

$$A_k = \int \chi_k(x_1)\chi_k(x_2)\chi_0(x_1')\chi_0(x_2')\ dx_1 dx_2. \qquad (7.23)$$

In order to carry out this calculation, let us use the generating function of the Hermite polynomials [14, 16],

$$G(r, z) = e^{(-r^2 + 2rz)} = \sum_m \frac{r^m}{m!} H_m(z), \qquad (7.24)$$

and evaluate the integral

$$I = \int G(r, x_1) G(s, x_2)$$
$$\times \exp\left\{-\left(\frac{x_1^2 + x_2^2}{2} + \frac{e^{-2\eta}(x_1 + x_2)^2 + e^{2\eta}(x_1 - x_2)^2}{4}\right)\right\} dx_1\ dx_2. \qquad (7.25)$$

The integrand becomes one exponential function, and its exponent is quadratic in x_1 and x_2. This quadratic form can be diagonalized, and the integral can be evaluated [2, 12]. The result is

$$I = \left[\frac{\pi}{\cosh\eta}\right]\exp\{(2rs\tanh\eta)\}\exp\left\{\left(\frac{2rs}{\cosh\eta}\right)\right\}. \qquad (7.26)$$

We can now Taylor-expand this expression and choose the coefficients of $(rs)^k$, for $H_k(x_1)$, $H_k(x_2)$, respectively. The result is

$$A_k = \left(\frac{1}{\cosh\eta}\right)(\tanh\eta)^k, \qquad (7.27)$$

with $\beta = \tanh\eta$. Thus, the series becomes

$$|\beta\rangle = \left(\frac{1}{\cosh\eta}\right)\sum_k (\tanh\eta)^k \chi_k(x_1)\chi_k(x_2). \tag{7.28}$$

This result leads to

$$|\beta\rangle = \exp\left(\beta\left[a_1^\dagger a_2^\dagger - a_1 a_2\right]\right)|0,0\rangle = \sqrt{1-\beta^2}\sum_k \beta^k |k,k\rangle. \tag{7.29}$$

7.3 Symmetries of two-mode states

In section 7.2 we studied in detail the role of

$$Q_3 = \frac{i}{2}(a_1^\dagger a_2^\dagger - a_1 a_2), \tag{7.30}$$

which leads to the generation of the two-mode coherent state or the squeezed state.

In their paper of 1986 [22], Yurke *et al* considered possible interferometers exhibiting the following two additional operators:

$$S_3 = \frac{1}{2}(a_1^\dagger a_1 + a_2 a_2^\dagger),$$
$$K_3 = \frac{1}{2}(a_1^\dagger a_2^\dagger + a_1 a_2). \tag{7.31}$$

The three Hermitian operators from equations (7.30) and (7.31) satisfy the commutation relations

$$[K_3, Q_3] = -iS_3, \quad [Q_3, S_3] = iK_3, \quad \text{and} \quad [S_3, K_3] = iQ_3. \tag{7.32}$$

These relations are like those for the $SU(1,1)$ group or the Lorentz group $O(2,1)$, applicable to two space dimensions and one time dimension, as we discussed in chapters 2 and 3.

In addition, in the same paper Yurke *et al* discussed the possibility of constructing interferometers exhibiting the symmetry generated by

$$L_1 = \frac{1}{2}(a_1^\dagger a_2 + a_2^\dagger a_1),$$
$$L_2 = \frac{1}{2i}(a_1^\dagger a_2 - a_2^\dagger a_1),$$
$$L_3 = \frac{1}{2}(a_1^\dagger a_1 - a_2^\dagger a_2). \tag{7.33}$$

These generators satisfy the closed set of commutation relations

$$[L_i, L_j] = i\epsilon_{ijk}L_k, \tag{7.34}$$

and therefore define a Lie algebra which is the same as that for the $SU(2)$ group or the three-dimensional rotation group.

We are then led to ask whether it is possible to construct a Lie algebra containing the six Hermitian operators from equations (7.30), (7.31), and (7.32). It is not possible. We have to add four additional operators, namely

$$K_1 = -\frac{1}{4}(a_1^\dagger a_1^\dagger + a_1 a_1 - a_2^\dagger a_2^\dagger - a_2 a_2),$$

$$K_2 = +\frac{i}{4}(a_1^\dagger a_1^\dagger - a_1 a_1 + a_2^\dagger a_2^\dagger - a_2 a_2),$$

$$Q_1 = -\frac{i}{4}(a_1^\dagger a_1^\dagger - a_1 a_1 - a_2^\dagger a_2^\dagger + a_2 a_2),$$

$$Q_2 = -\frac{1}{4}(a_1^\dagger a_1^\dagger + a_1 a_1 + a_2^\dagger a_2^\dagger + a_2 a_2).$$

$$(7.35)$$

There are now ten operators from equations (7.30)–(7.33) and (7.35). Indeed, these ten operators satisfy the following closed set of commutation relations:

$$[L_i, L_j] = i\epsilon_{ijk}L_k, \qquad [L_i, K_j] = i\epsilon_{ijk}K_k, \qquad [L_i, Q_j] = i\epsilon_{ijk}Q_k,$$
$$[K_i, K_j] = [Q_i, Q_j] = -i\epsilon_{ijk}L_k, \qquad [L_i, S_3] = 0,$$
$$[K_i, Q_j] = -i\delta_{ij}S_3, \qquad [K_i, S_3] = -iQ_i, \qquad [Q_i, S_3] = iK_i.$$

$$(7.36)$$

As Dirac noted in 1963 [5], this set is the same as the Lie algebra for $O(3, 2)$, the de Sitter group, with ten generators. This group is the Lorentz group applicable to the three-dimensional space with two time variables. This aspect of the space–time symmetry was discussed in [3] and in chapter 1. In this very paper Dirac also mentioned that this set of commutation relations serves as the Lie algebra for the four-dimensional symplectic group commonly called $Sp(4)$, applicable to the systems of two one-dimensional particles, each with a two-dimensional phase space.

For a dynamical system consisting of two pairs of canonical variables x_1, p_1 and x_2, p_2, we can use the four-dimensional space with the coordinate variables defined as [8]

$$(x_1, p_1, x_2, p_2).$$

$$(7.37)$$

Then the four-by-four transformation matrix M applicable to this four-component vector is canonical if [1, 7]

$$M J \tilde{M} = J,$$

$$(7.38)$$

where $\tilde{M}$ is the transpose of the M matrix. Here

$$J = \begin{pmatrix} 0 & 1 & 0 & 0 \\ -1 & 0 & 0 & 0 \\ 0 & 0 & 0 & 1 \\ 0 & 0 & -1 & 0 \end{pmatrix}.$$

$$(7.39)$$

According to this form of the J matrix, the area of the phase space for the x_1 and p_1 variables remains invariant, and the story is the same for the phase space of x_2 and p_2.

We can then write the generators of the $Sp(4)$ group as

$$L_1 = -\frac{1}{2}\begin{pmatrix} 0 & I \\ I & 0 \end{pmatrix}\sigma_2, \qquad L_2 = \frac{i}{2}\begin{pmatrix} 0 & -I \\ I & 0 \end{pmatrix}I, \qquad (7.40)$$

$$L_3 = \frac{1}{2}\begin{pmatrix} -I & 0 \\ 0 & I \end{pmatrix}\sigma_2, \qquad S_3 = \frac{1}{2}\begin{pmatrix} I & 0 \\ 0 & I \end{pmatrix}\sigma_2, \qquad (7.41)$$

and

$$K_1 = \frac{i}{2}\begin{pmatrix} I & 0 \\ 0 & -I \end{pmatrix}\sigma_1, \quad K_2 = \frac{i}{2}\begin{pmatrix} I & 0 \\ 0 & I \end{pmatrix}\sigma_3, \quad K_3 = -\frac{i}{2}\begin{pmatrix} 0 & I \\ I & 0 \end{pmatrix}\sigma_1, \qquad (7.42)$$

$$Q_1 = -\frac{i}{2}\begin{pmatrix} I & 0 \\ 0 & -I \end{pmatrix}\sigma_3, \quad Q_2 = \frac{i}{2}\begin{pmatrix} I & 0 \\ 0 & I \end{pmatrix}\sigma_1, \quad Q_3 = \frac{i}{2}\begin{pmatrix} 0 & I \\ I & 0 \end{pmatrix}\sigma_3, \qquad (7.43)$$

where I is the two-by-two identity matrix, while σ_1, σ_2, and σ_3 are the two-by-two Pauli matrices.

These four-by-four matrices satisfy the commutation relations given in equation (7.36). Indeed, the group generated by these four-by-four matrices is like (locally isomorphic to) the group of five-by-five matrices for the $O(3, 2)$ de Sitter group.

There are fifteen four-by-four Dirac matrices. Only ten of those matrices serve as the generators of the $Sp(4)$ group. Then the question arises as to what role the remaining five matrices play in the four-dimensional phase space. We will resolve this issue in the following sections.

7.4 Dirac matrices and $O(3,3)$ symmetry

We are dealing here with the four-by-four transformation matrices, with real elements. This means that all elements of the generators are imaginary. Thus, we choose the Dirac matrices in the Majorana representation [15].

Since all the generators for the two coupled oscillator system can be written as four-by-four matrices with imaginary elements, it is convenient to work with Dirac matrices in the Majorana representation [9, 11, 13, 15]. In the Majorana representation, the four γ matrices for the Dirac equation are

$$\begin{aligned} \gamma_1 &= i\begin{pmatrix} I & 0 \\ 0 & I \end{pmatrix}\sigma_3, & \gamma_2 &= \begin{pmatrix} 0 & -I \\ I & 0 \end{pmatrix}\sigma_2, \\ \gamma_3 &= -i\begin{pmatrix} I & 0 \\ 0 & I \end{pmatrix}\sigma_1, & \gamma_0 &= \begin{pmatrix} 0 & I \\ I & 0 \end{pmatrix}\sigma_2. \end{aligned} \qquad (7.44)$$

These γ matrices are transformed like four-vectors under Lorentz transformations. From these four matrices, we can construct one pseudo-scalar matrix

$$\gamma_5 = i\gamma_0\gamma_1\gamma_2\gamma_3 = \begin{pmatrix} I & 0 \\ 0 & -I \end{pmatrix}\sigma_2, \tag{7.45}$$

and a pseudo-vector $i\gamma_5\gamma_\mu$ consisting of

$$i\gamma_5\gamma_1 = -i\begin{pmatrix} I & 0 \\ 0 & -I \end{pmatrix}\sigma_1, \qquad i\gamma_5\gamma_2 = -i\begin{pmatrix} 0 & I \\ I & 0 \end{pmatrix}I,$$

$$i\gamma_5\gamma_0 = i\begin{pmatrix} 0 & I \\ -I & 0 \end{pmatrix}I, \qquad i\gamma_5\gamma_3 = -i\begin{pmatrix} I & 0 \\ 0 & -I \end{pmatrix}\sigma_3. \tag{7.46}$$

In addition, we can construct the tensor of γ as

$$T_{\mu\nu} = \frac{i}{2}(\gamma_\mu\gamma_\nu - \gamma_\nu\gamma_\mu). \tag{7.47}$$

This anti-symmetric tensor has six components. They are

$$i\gamma_0\gamma_1 = -i\begin{pmatrix} 0 & I \\ I & 0 \end{pmatrix}\sigma_1, \quad i\gamma_0\gamma_2 = i\begin{pmatrix} I & 0 \\ 0 & -I \end{pmatrix}I, \quad i\gamma_0\gamma_3 = -i\begin{pmatrix} 0 & I \\ I & 0 \end{pmatrix}\sigma_3, \tag{7.48}$$

and

$$i\gamma_1\gamma_2 = i\begin{pmatrix} 0 & -I \\ I & 0 \end{pmatrix}\sigma_1, \quad i\gamma_2\gamma_3 = -i\begin{pmatrix} 0 & -I \\ I & 0 \end{pmatrix}\sigma_3, \quad i\gamma_3\gamma_1 = \begin{pmatrix} I & 0 \\ 0 & I \end{pmatrix}\sigma_2. \tag{7.49}$$

There are now fifteen linearly independent four-by-four traceless matrices with imaginary components [13]. These are the fifteen Dirac matrices in the Majorana representation. They are given in table 7.1.

The four-by-four expressions for the L_i, K_i, Q_i are given in section 7.3. The expression for S_3 is also given there.

According to table 7.1, S_1 and S_2 can be written as

$$S_1 = \frac{i}{2}\begin{pmatrix} 0 & I \\ -I & 0 \end{pmatrix}\sigma_3 \quad \text{and} \quad S_2 = \frac{i}{2}\begin{pmatrix} 0 & -I \\ I & 0 \end{pmatrix}\sigma_1, \tag{7.50}$$

and G_i as

$$G_1 = -\frac{i}{2}\begin{pmatrix} 0 & I \\ I & 0 \end{pmatrix}I, \quad G_2 = \frac{1}{2}\begin{pmatrix} 0 & I \\ I & 0 \end{pmatrix}\sigma_2, \quad \text{and} \quad G_3 = \frac{i}{2}\begin{pmatrix} I & 0 \\ 0 & -I \end{pmatrix}I. \tag{7.51}$$

Unlike the ten generators given in section 7.3, these five generators do not lead to canonical transformations. They lead to transformations which can increase the area of phase space for one mode while decreasing the area of the other. We shall study this aspect of the uncertainty relations in more detail using the Wigner function.

If these five generators are added to those ten canonical generators, there are fifteen generators as given in table 7.1. Indeed, these fifteen generators satisfy the following set of commutation relations:

Table 7.1. Fifteen Dirac matrices in the Majorana representation applicable to the four-dimensional phase space. From the $O(3, 3)$ point of view with three space-like and three time-like dimensions, there are two sets of rotation generators and three sets of three squeeze generators.

	First component	Second component	Third component
Rotation	$L_1 = -\frac{i}{2}\gamma_0$	$L_2 = -\frac{i}{2}\gamma_5\gamma_0$	$L_3 = -\frac{1}{2}\gamma_5$
Rotation	$S_1 = \frac{i}{2}\gamma_2\gamma_3$	$S_2 = \frac{i}{2}\gamma_1\gamma_2$	$S_3 = \frac{i}{2}\gamma_3\gamma_1$
Boost	$K_1 = -\frac{i}{2}\gamma_5\gamma_1$	$K_2 = \frac{1}{2}\gamma_1$	$K_3 = \frac{i}{2}\gamma_0\gamma_1$
Boost	$Q_1 = \frac{i}{2}\gamma_5\gamma_3$	$Q_2 = -\frac{1}{2}\gamma_3$	$Q_3 = -\frac{i}{2}\gamma_0\gamma_3$
Boost	$G_1 = -\frac{i}{2}\gamma_5\gamma_2$	$G_2 = \frac{1}{2}\gamma_2$	$G_3 = \frac{i}{2}\gamma_0\gamma_2$

$$
\begin{aligned}
&[L_i, L_j] = i\epsilon_{ijk}L_k, \qquad [S_i, S_j] = i\epsilon_{ijk}S_k, \qquad [L_i, S_j] = 0, \\
&[L_i, K_j] = i\epsilon_{ijk}K_k, \qquad [L_i, Q_j] = i\epsilon_{ijk}Q_k, \qquad [L_i, G_j] = i\epsilon_{ijk}G_k, \\
&[K_i, K_j] = [Q_i, Q_j] = [G_i, G_j] = -i\epsilon_{ijk}L_k, \\
&[K_i, Q_j] = -i\delta_{ij}S_3, \qquad [Q_i, G_j] = -i\delta_{ij}S_1, \qquad [G_i, K_j] = -i\delta_{ij}S_2, \\
&[K_i, S_3] = -iQ_i, \qquad [Q_i, S_3] = iK_i, \qquad [G_i, S_3] = 0, \\
&[K_i, S_1] = 0, \qquad [Q_i, S_1] = -iG_i, \qquad [G_i, S_1] = iQ_i, \\
&[K_i, S_2] = iG_i, \qquad [Q_i, S_2] = 0, \qquad [G_i, S_2] = -iK_i.
\end{aligned}
\tag{7.52}
$$

This set of commutation relations serves as the Lie algebra for the group $SL(4, r)$ or the group of four-by-four unimodular matrices with real elements. This set can also serve as the Lie algebra for the $O(3, 3)$ Lorentz group.

As can be seen from table 7.1, there are six anti-symmetric, therefore Hermitian, and nine symmetric, non-Hermitian, matrices. These anti-symmetric matrices were divided into two sets of three rotation generators in the four-dimensional phase space. The nine symmetric matrices can be divided into three sets of three squeeze generators. However, this classification scheme is easier to understand in terms of the group $O(3, 3)$.

7.5 Symmetries in phase space

In order to thoroughly study the symmetry problems of the two-mode state, we need the Wigner phase space representation of quantum mechanics [10, 20]. From chapter 5, the Wigner phase-space distribution function is defined as

$$
W(x_1, x_2, p_1, p_2) = \left(\frac{1}{\pi}\right)^2 \int e^{2i(p_1 y_1 + p_2 y_2)}
$$
$$
\times \rho(x_1 - y_1, x_2 - y_2; x_1 + y_1, x_2 + y_2) \; dy_1 \; dy_2,
\tag{7.53}
$$

while the density matrix is defined as [18]

$$\rho(x_1, x_2; x_1', x_2') = \psi(x_1, x_2)\psi^*(x_1', x_2').$$ (7.54)

If the x_2 variable is not observed, this density matrix becomes

$$\rho(x_1, x_1') = \int \rho(x_1, x_2; x_1', x_2) \, dx_2,$$ (7.55)

and the trace of this density matrix is 1:

$$\int \rho(x, x)dx = 1.$$ (7.56)

The vacuum state of the two-mode state corresponds to the ground states of the two-oscillator system with the wave function

$$\psi(x_1, x_2) = \frac{1}{\sqrt{\pi}} \exp\left\{-\frac{1}{2}\left(x_1^2 + x_2^2\right)\right\}.$$ (7.57)

Thus the density matrix is

$$\rho(x_1, x_2; x_1', x_2') = \frac{1}{\pi} \exp\left\{-\frac{1}{2}\left(x_1^2 + x_2^2 + (x_1')^2 + (x_2')^2\right)\right\}.$$ (7.58)

Thus, the Wigner function for the ground state is

$$W(x_1, x_2, p_1, p_2) = \left(\frac{1}{\pi}\right)^2 \exp\left\{-\left(x_1^2 + x_2^2 + p_1^2 + p_2^2\right)\right\}.$$ (7.59)

This Wigner function is defined in the four-dimensional phase space where the transformations defined in section 7.4 are applicable. According to the fifteen four-by-four Dirac matrices, there are fifteen generators of the transformations.

Since Dirac matrices are constructed for the description of Einstein's special relativity, we have, in section 7.4, organized those operators according to the representation of the Lorentz group, namely the Lorentz group applicable to the three space and three time variables. The size of the transformation matrices are six-by-six. This six-by-six representation is beyond the scope of this book.

In this section, we use the Wigner function to organize those generators according to the four-dimensional phase space. First of all, we can write these fifteen generators in the following way and consider the following fifteen traceless matrices.

Let us first examine the $O(3, 3)$ content of the Wigner function. There are four possible rotation generators, namely

$$L_1 = +\left(\frac{i}{2}\right)\left\{\left(x_1\frac{\partial}{\partial p_2} - p_2\frac{\partial}{\partial x_1}\right) + \left(x_2\frac{\partial}{\partial p_1} - p_1\frac{\partial}{\partial x_2}\right)\right\},$$

$$L_2 = -\left(\frac{i}{2}\right)\left\{\left(x_1\frac{\partial}{\partial x_2} - x_2\frac{\partial}{\partial x_1}\right) + \left(p_1\frac{\partial}{\partial p_2} - p_2\frac{\partial}{\partial p_1}\right)\right\},$$

$$L_3 = +\left(\frac{i}{2}\right)\left\{\left(x_1\frac{\partial}{\partial p_1} - p_1\frac{\partial}{\partial x_1}\right) - \left(x_2\frac{\partial}{\partial p_2} - p_2\frac{\partial}{\partial x_2}\right)\right\},$$

$$S_3 = +\left(\frac{i}{2}\right)\left\{\left(x_1\frac{\partial}{\partial p_1} - p_1\frac{\partial}{\partial x_1}\right) + \left(x_2\frac{\partial}{\partial p_2} - p_2\frac{\partial}{\partial x_2}\right)\right\}.$$

(7.60)

It is noted that S_3 and L_3 generate rotations in the (x_1, p_1) and (x_2, p_2) spaces. They do not mix up the the first and second phase spaces. The rotations are in the same direction for S_3 and in the opposite directions in the case of L_3. Unlike L_3 or S_3, L_1 and L_2 generate rotations in (x_1, p_2) and (x_2, p_1), respectively. They mix up the phase spaces.

The first set of three squeeze generators is

$$K_1 = -\left(\frac{i}{2}\right)\left\{\left(x_1\frac{\partial}{\partial p_1} + p_1\frac{\partial}{\partial x_1}\right) - \left(x_2\frac{\partial}{\partial p_2} + p_2\frac{\partial}{\partial x_2}\right)\right\},$$

$$K_2 = -\left(\frac{i}{2}\right)\left\{\left(x_1\frac{\partial}{\partial x_1} - p_2\frac{\partial}{\partial p_2}\right) + \left(x_2\frac{\partial}{\partial x_2} - p_1\frac{\partial}{\partial p_1}\right)\right\},$$

$$K_3 = +\left(\frac{i}{2}\right)\left\{\left(x_1\frac{\partial}{\partial p_2} + p_2\frac{\partial}{\partial x_1}\right) + \left(x_2\frac{\partial}{\partial p_1} + p_1\frac{\partial}{\partial x_2}\right)\right\}.$$

(7.61)

It is clear that L_1, K_2, and K_3 will satisfy the set of commutation relations for $(2 + 1)$-like transformations in the (x_1, p_2) and (x_2, p_1) spaces separately. Likewise, the subgroups generated by L_2, K_3, K_1 and by L_3, K_1, K_2 perform $(2 + 1)$-like transformations in two separate two-dimensional phase spaces.

The second set of squeeze generators is

$$Q_1 = +\left(\frac{i}{2}\right)\left\{\left(x_1\frac{\partial}{\partial x_1} - p_1\frac{\partial}{\partial p_1}\right) - \left(x_2\frac{\partial}{\partial x_2} - p_2\frac{\partial}{\partial p_2}\right)\right\},$$

$$Q_2 = -\left(\frac{i}{2}\right)\left\{\left(x_1\frac{\partial}{\partial p_1} + p_1\frac{\partial}{\partial x_1}\right) + \left(x_2\frac{\partial}{\partial p_2} + p_2\frac{\partial}{\partial x_2}\right)\right\},$$

$$Q_3 = -\left(\frac{i}{2}\right)\left\{\left(x_2\frac{\partial}{\partial x_1} + x_1\frac{\partial}{\partial x_2}\right) - \left(p_2\frac{\partial}{\partial p_1} + p_1\frac{\partial}{\partial p_2}\right)\right\}.$$

(7.62)

The ten generators given in equations (7.60), (7.61), and (7.62) satisfy the set of commutation relations given in equation (7.36), known as the Lie algebra for $O(3, 2)$ or $Sp(4)$. We tabulate the ten generators of canonical transformations in table 7.2.

Table 7.2. Ten four-by-four matrices applicable to the four-dimensional phase space. They are organized according to the Pauli matrices. The four-by-four matrices here are expressed in terms of the two-by-two matrices of two-by-two Pauli matrices. Fortunately, these two sets are factorizable, and they can be written as given in this table.

using	σ_3	σ_1	σ_2
Dirac	$A_1 = \gamma_1$	$A_2 = -\gamma_3$	$A_3 = i\gamma_3\gamma_1$
$O(3,2)$	K_2	Q_2	S_3
$\begin{pmatrix} I & 0 \\ 0 & I \end{pmatrix}$	$\frac{i}{2}\begin{pmatrix} I & 0 \\ 0 & I \end{pmatrix}\sigma_3$	$\frac{i}{2}\begin{pmatrix} I & 0 \\ 0 & I \end{pmatrix}\sigma_1$	$\frac{1}{2}\begin{pmatrix} I & 0 \\ 0 & I \end{pmatrix}\sigma_2$
Dirac	$B_1 = i\gamma_3\gamma_5$	$B_2 = i\gamma_1\gamma_5$	$B_3 = i\gamma_5$
$O(3,2)$	Q_1	K_1	L_3
$\begin{pmatrix} I & 0 \\ 0 & -I \end{pmatrix}$	$\frac{i}{2}\begin{pmatrix} I & 0 \\ 0 & -I \end{pmatrix}\sigma_3$	$\frac{i}{2}\begin{pmatrix} I & 0 \\ 0 & -I \end{pmatrix}\sigma_1$	$\frac{1}{2}\begin{pmatrix} I & 0 \\ 0 & -I \end{pmatrix}\sigma_2$
Dirac	$C_1 = i\gamma_0\gamma_3$	$C_2 = -i\gamma_0\gamma_1$	$C_3 = \gamma_0$
$O(3,2)$	Q_3	$-K_3$	$-L_1$
$\begin{pmatrix} 0 & I \\ I & 0 \end{pmatrix}$	$\frac{i}{2}\begin{pmatrix} 0 & I \\ I & 0 \end{pmatrix}\sigma_3$	$\frac{i}{2}\begin{pmatrix} 0 & I \\ I & 0 \end{pmatrix}\sigma_1$	$\frac{1}{2}\begin{pmatrix} 0 & I \\ I & 0 \end{pmatrix}\sigma_2$
Dirac	$D_0 = -i\gamma_0\gamma_5$		
$O(3,2)$	L_2		
$\begin{pmatrix} 0 & -I \\ I & 0 \end{pmatrix}$	$\frac{i}{2}\begin{pmatrix} 0 & -I \\ I & 0 \end{pmatrix}I$		

According to equation (7.50), there are two additional rotation generators which can be written as

$$S_1 = +\frac{i}{2}\left\{\left(x_1\frac{\partial}{\partial x_2} - x_2\frac{\partial}{\partial x_1}\right) - \left(p_1\frac{\partial}{\partial p_2} - p_2\frac{\partial}{\partial p_1}\right)\right\},$$
$$S_2 = -\frac{i}{2}\left\{\left(x_1\frac{\partial}{\partial p_2} - p_2\frac{\partial}{\partial x_1}\right) + \left(x_2\frac{\partial}{\partial p_1} - p_1\frac{\partial}{\partial x_2}\right)\right\}. \tag{7.63}$$

In addition, according to equation (7.51), there are three additional squeeze operators. They can be written as

$$G_1 = -\frac{i}{2}\left\{\left(x_1\frac{\partial}{\partial x_2} + x_2\frac{\partial}{\partial x_1}\right) + \left(p_1\frac{\partial}{\partial p_2} + p_2\frac{\partial}{\partial p_1}\right)\right\},$$
$$G_2 = +\frac{i}{2}\left\{\left(x_1\frac{\partial}{\partial p_2} + p_2\frac{\partial}{\partial x_1}\right) - \left(x_2\frac{\partial}{\partial p_1} + p_1\frac{\partial}{\partial x_2}\right)\right\}, \tag{7.64}$$
$$G_3 = -\frac{i}{2}\left\{\left(x_1\frac{\partial}{\partial x_1} + p_1\frac{\partial}{\partial p_1}\right) + \left(x_2\frac{\partial}{\partial p_2} + p_2\frac{\partial}{\partial x_2}\right)\right\}.$$

These five additional generators lead to the fifteen generators for the $O(3, 3)$ group as discussed in section 7.4. They also correspond to fifteen four-by-four Dirac matrices. These four-by-four matrices are tabulated in table 7.3.

We can write these generators in terms of the four-by-four matrices applicable to the four-dimensional phase space (x_1, p_1, x_2, p_2). Of the ten generators, nine of them are placed in the three rows and three columns in that table, and one of them does not belong to this three-by-three table.

In table 7.2 for the generators of canonical transformations the rows are specified by the four-by-four matrices

$$A = \begin{pmatrix} I & 0 \\ 0 & I \end{pmatrix}, \quad B = \begin{pmatrix} I & 0 \\ 0 & -I \end{pmatrix}, \quad \text{and} \quad C = \begin{pmatrix} 0 & I \\ I & 0 \end{pmatrix}, \tag{7.65}$$

where I is the two-by-two unit matrix.

The columns are specified by σ_3, σ_1, and σ_2. If we let

$$K_1 = \frac{i}{2}\sigma_1, \quad K_3 = \frac{i}{2}\sigma_3, \quad \text{and} \quad L_2 = \frac{1}{2}\sigma_2, \tag{7.66}$$

Table 7.3. Fifteen Dirac matrices in the Majorana representation applicable to the four-dimensional phase space. There are sixteen items in this table, but A_0 should be excluded, because it is an identity matrix whose trace does not vanish. These matrices serve also as the generators of the $O(3, 3)$ Lorentz group applicable to three space-like and three time-like dimensions. They are systematically tabulated in table 7.1.

using	I	σ_3	σ_1	σ_2
Dirac	$A_0 = iI$	$A_1 = \gamma_1$	$A_2 = -\gamma_3$	$A_3 = i\gamma_3\gamma_1$
$O(3, 3)$		K_2	Q_2	S_3
$\begin{pmatrix} I & 0 \\ 0 & I \end{pmatrix}$	$\frac{i}{2}\begin{pmatrix} I & 0 \\ 0 & I \end{pmatrix}I$	$\frac{i}{2}\begin{pmatrix} I & 0 \\ 0 & I \end{pmatrix}\sigma_3$	$\frac{i}{2}\begin{pmatrix} I & 0 \\ 0 & I \end{pmatrix}\sigma_1$	$\frac{1}{2}\begin{pmatrix} I & 0 \\ 0 & I \end{pmatrix}\sigma_2$
Dirac	$B_0 = i\gamma_0\gamma_2$	$B_1 = i\gamma_3\gamma_5$	$B_2 = i\gamma_1\gamma_5$	$B_3 = i\gamma_5$
$O(3, 3)$	G_3	Q_1	K_1	L_3
$\begin{pmatrix} I & 0 \\ 0 & -I \end{pmatrix}$	$\frac{i}{2}\begin{pmatrix} I & 0 \\ 0 & -I \end{pmatrix}I$	$\frac{1}{2}\begin{pmatrix} I & 0 \\ 0 & -I \end{pmatrix}\sigma_3$	$\frac{i}{2}\begin{pmatrix} I & 0 \\ 0 & -I \end{pmatrix}\sigma_1$	$\frac{1}{2}\begin{pmatrix} I & 0 \\ 0 & -I \end{pmatrix}\sigma_2$
Dirac	$C_0 = i\gamma_2\gamma_5$	$C_1 = i\gamma_0\gamma_3$	$C_2 = -i\gamma_0\gamma_1$	$C_3 = \gamma_0$
$O(3, 3)$	$-G_1$	Q_3	$-K_3$	$-L_1$
$\begin{pmatrix} 0 & I \\ I & 0 \end{pmatrix}$	$\frac{i}{2}\begin{pmatrix} 0 & I \\ I & 0 \end{pmatrix}I$	$\frac{i}{2}\begin{pmatrix} 0 & I \\ I & 0 \end{pmatrix}\sigma_3$	$\frac{i}{2}\begin{pmatrix} 0 & I \\ I & 0 \end{pmatrix}\sigma_1$	$\frac{1}{2}\begin{pmatrix} 0 & I \\ I & 0 \end{pmatrix}\sigma_2$
Dirac	$D_0 = -i\gamma_0\gamma_5$	$D_1 = -i\gamma_2\gamma_3$	$D_2 = i\gamma_1\gamma_2$	$D_3 = i\gamma_2$
$O(3, 3)$	L_2	S_1	S_2	G_2
$\begin{pmatrix} 0 & -I \\ I & 0 \end{pmatrix}$	$\frac{i}{2}\begin{pmatrix} 0 & -I \\ I & 0 \end{pmatrix}I$	$\frac{i}{2}\begin{pmatrix} 0 & -I \\ I & 0 \end{pmatrix}\sigma_3$	$\frac{i}{2}\begin{pmatrix} 0 & -I \\ I & 0 \end{pmatrix}\sigma_1$	$\frac{1}{2}\begin{pmatrix} 0 & -I \\ I & 0 \end{pmatrix}\sigma_2$

we end up the following closed set of commutation relations:

$$[K_1, K_3] = iL_2, \quad [L_2, K_3] = iK_1, \quad \text{and} \quad [L_2, K_1] = -iK_3. \tag{7.67}$$

This set is the Lie algebra for the $Sp(2)$ group. Here all the elements are imaginary numbers, and the transformation matrices consist of real elements. Thus, it is a subgroup of the six-parameter $SL(2, c)$ group.

As was noted in chapter 2, the $Sp(2)$ group forms a Lie algebra where all the generators are imaginary and all the transform matrices are real. Therefore this group has many applications in the squeezed states of light. Thus, the columns are organized according to the $Sp(2)$ group.

In order to produce the ten generators for canonical transformations, we need one additional element, given also in table 7.2. The generator L_2 does not belong to the three-by-three table, but it can produce its own row and column, as shown in table 7.3. This table contains one additional row with the common matrix

$$D = \begin{pmatrix} 0 & I \\ -I & 0 \end{pmatrix}. \tag{7.68}$$

This fourth row contains S_1, S_2, G_2, which can also serve as the $Sp(2)$ Lie algebra.

In table 7.3, the first column is added to table 7.2, consisting of the $B_0 = G_3$, $C_0 = -G_1$, and $D_0 = L_2$ matrices, namely

$$i\begin{pmatrix} I & 0 \\ 0 & -I \end{pmatrix}, \quad i\begin{pmatrix} 0 & I \\ I & 0 \end{pmatrix}, \quad \text{and} \quad i\begin{pmatrix} 0 & -I \\ I & 0 \end{pmatrix}. \tag{7.69}$$

These matrices also constitute the Lie algebra for the $Sp(2)$ group.

It is interesting to note that the symmetry of the two-oscillator system consists of the symmetries of the $Sp(2)$ symmetries.

We can summarize the distributions of the $Sp(2)$ symmetries with the following two traceless matrices:

$$\begin{pmatrix} A_1 & A_2 & A_3 \\ B_1 & B_2 & B_3 \\ C_1 & C_2 & C_3 \\ D_1 & D_2 & D_3 \end{pmatrix}, \quad \text{and} \quad \begin{pmatrix} B_0 & B_1 & B_2 & B_3 \\ C_0 & C_1 & C_2 & C_3 \\ D_0 & D_1 & D_2 & D_3 \end{pmatrix}. \tag{7.70}$$

The $Sp(2)$ symmetry is seen in each row in the first matrix and in each column in the second matrix.

7.6 Two coupled oscillators

We are familiar with the physics of coupled oscillators. With appropriate coordinate transformations, the total Hamiltonians can be separated into that of the two independent systems. As a consequence, all the transformation matrices can become diagonal.

Among the four-by-four matrices given in table 7.3, B_0 and B_1 generate diagonal transformation matrices applicable to the four-component vector (x_1, p_1, x_2, p_2). The transformation matrices are

$$T_0 = e^{-i\eta B_0} = \begin{pmatrix} e^\eta & 0 & 0 & 0 \\ 0 & e^\eta & 0 & 0 \\ 0 & 0 & e^{-\eta} & 0 \\ 0 & 0 & 0 & e^{-\eta} \end{pmatrix},$$

$$T_1 = e^{-i\eta B_1} = \begin{pmatrix} e^\eta & 0 & 0 & 0 \\ 0 & e^{-\eta} & 0 & 0 \\ 0 & 0 & e^{-\eta} & 0 \\ 0 & 0 & 0 & e^\eta \end{pmatrix},$$

(7.71)

respectively. The four-by-four T_0 and T_1 matrices perform non-canonical and canonical transformations, respectively.

Let us examine the symmetries derived in section 7.5 in terms of two harmonic oscillators. The Hamiltonian for each oscillator can be written as

$$H_i = \frac{1}{2}\left(x_i^2 + p_i^2\right).$$

(7.72)

Then the total Hamiltonian becomes

$$H_+ = H_1 + H_2 = \frac{1}{2}\left(x_1^2 + x_2^2 + p_1^2 + p_2^2\right).$$

(7.73)

In addition, in view of equation (7.20), we can consider also

$$H_- = H_1 - H_2 = \frac{1}{2}\left(x_1^2 - x_2^2 + p_1^2 - p_2^2\right).$$

(7.74)

This Hamiltonian produces the same set of wave functions as the Hamiltonian H_+.

While the Hamiltonian H_+ is invariant under rotations,

$$\begin{pmatrix} x_1' \\ x_2' \end{pmatrix} = \begin{pmatrix} \cos\alpha & \sin\alpha \\ -\sin\alpha & \cos\alpha \end{pmatrix}\begin{pmatrix} x_1 \\ x_2 \end{pmatrix},$$

$$\begin{pmatrix} p_1' \\ p_2' \end{pmatrix} = \begin{pmatrix} \cos\alpha & \pm\sin\alpha \\ \mp\sin\alpha & \cos\alpha \end{pmatrix}\begin{pmatrix} p_1 \\ p_2 \end{pmatrix},$$

(7.75)

H_- is invariant under the squeeze transformations [6],

$$\begin{pmatrix} x_1' \\ x_2' \end{pmatrix} = \begin{pmatrix} \cosh\eta & \sinh\eta \\ \sinh\eta & \cosh\eta \end{pmatrix}\begin{pmatrix} x_1 \\ x_2 \end{pmatrix},$$

$$\begin{pmatrix} p_1' \\ p_2' \end{pmatrix} = \begin{pmatrix} \cosh\eta & \pm\sinh\eta \\ \pm\sinh\eta & \cosh\eta \end{pmatrix}\begin{pmatrix} p_1 \\ p_2 \end{pmatrix}.$$

(7.76)

The ground-state Wigner function is

$$W(0) = \left(\frac{1}{\pi}\right)^2 \exp\left\{-\left(x_1^2 + x_2^2 + p_1^2 + p_2^2\right)\right\}. \tag{7.77}$$

Under the canonical transformation of $T_1(\eta)$, this Wigner function becomes

$$W_{c,\eta} = \left(\frac{1}{\pi}\right)^2 \exp\left\{-\left(e^{-2\eta}x_1^2 + e^{2\eta}p_1^2 + e^{2\eta}x_2^2 + e^{-2\eta}p_2^2\right)\right\}. \tag{7.78}$$

This canonical transformation is illustrated in figure 7.2. The area of phase space is preserved for each mode.

As for the non-canonical transformation of $T_0(\eta)$, the Wigner function becomes [4, 10, 11]

$$W_{n,\eta} = \left(\frac{1}{\pi}\right)^2 \exp\left\{-\left(e^{-2\eta}x_1^2 + e^{-2\eta}p_1^2 + e^{2\eta}x_2^2 + e^{2\eta}p_2^2\right)\right\}. \tag{7.79}$$

As is illustrated also in figure 7.2, the area of the first mode expands while the other shrinks. The product of these two areas remain constant.

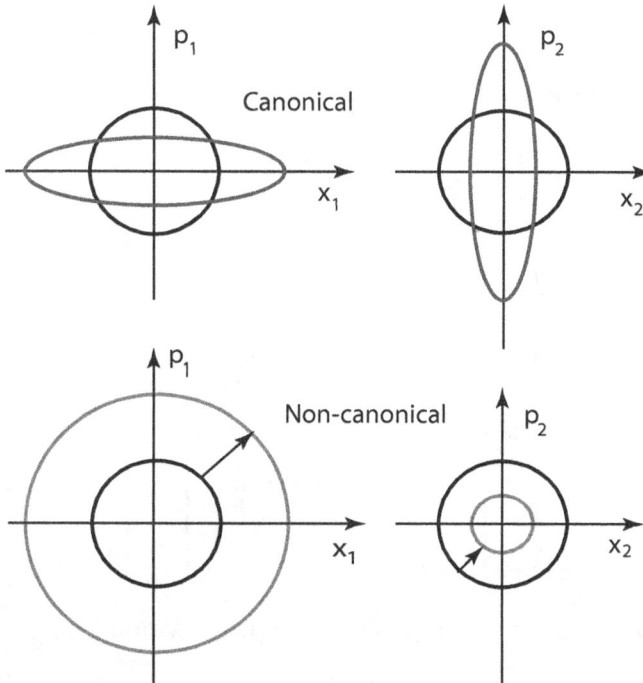

Figure 7.2. Canonical and non-canonical transformations. The area of each phase space is preserved under the canonical transformation. Under the non-canonical transformation, one of the areas expands and the other shrinks, while the product of these two areas remains invariant. This increase or decrease is in the statistical uncertainty. Thus, the decrease in uncertainty below the quantum limit does not violate the uncertainty principle.

It is more convenient to look at this system using the normal coordinate system, namely

$$x_\pm = \frac{1}{\sqrt{2}}(x_1 \pm x_2) \qquad \text{and} \qquad p_\pm = \frac{1}{\sqrt{2}}(p_1 \pm p_2). \tag{7.80}$$

These transformations can be expressed as

$$\begin{pmatrix} x_+ \\ p_+ \\ x_- \\ p_- \end{pmatrix} = \frac{1}{\sqrt{2}} \begin{pmatrix} 1 & 0 & 1 & 0 \\ 0 & 1 & 0 & 1 \\ 1 & 0 & -1 & 0 \\ 0 & 1 & 0 & -1 \end{pmatrix} \begin{pmatrix} x_1 \\ p_1 \\ x_2 \\ p_2 \end{pmatrix}. \tag{7.81}$$

This four-by-four matrix takes the form

$$\begin{pmatrix} I & I \\ I & -I \end{pmatrix} I = \begin{pmatrix} I & 0 \\ 0 & -I \end{pmatrix} I + \begin{pmatrix} 0 & I \\ I & 0 \end{pmatrix} I. \tag{7.82}$$

This four-by-four matrix commutes with the J matrix given in equation (7.39), which can be written as

$$J = i \begin{pmatrix} I & 0 \\ 0 & I \end{pmatrix} \sigma_2. \tag{7.83}$$

Thus the canonical condition for the phase spaces of (x_1, p_1) and (x_2, p_2) remains the same for the phase spaces of (x_+, p_+) and (x_-, p_-) .

In terms of these variables, the Hamiltonians take the form

$$H_+ = \frac{1}{2}\left(x_+^2 + x_-^2 + p_+^2 + p_-^2\right) \tag{7.84}$$

and

$$H_- = \frac{1}{2}(x_+ x_- + p_+ p_-). \tag{7.85}$$

The Hamiltonian H_+ is invariant under rotations, as in the case of equation (7.73). As for H_-, this Hamiltonian remains invariant under the transformation

$$\begin{pmatrix} x'_+ \\ p'_+ \\ x'_- \\ p'_- \end{pmatrix} = \begin{pmatrix} e^\eta & 0 & 0 & 0 \\ 0 & e^{-\eta} & 0 & 0 \\ 0 & 0 & e^{-\eta} & 0 \\ 0 & 0 & 0 & e^\eta \end{pmatrix} \begin{pmatrix} x_+ \\ p_+ \\ x_- \\ p_- \end{pmatrix}. \tag{7.86}$$

Furthermore, H_- is also invariant under the transformation

$$\begin{pmatrix} x'_+ \\ p'_+ \\ x'_- \\ p'_- \end{pmatrix} = \begin{pmatrix} e^\eta & 0 & 0 & 0 \\ 0 & e^\eta & 0 & 0 \\ 0 & 0 & e^{-\eta} & 0 \\ 0 & 0 & 0 & e^{-\eta} \end{pmatrix} \begin{pmatrix} x_+ \\ p_+ \\ x_- \\ p_- \end{pmatrix}. \tag{7.87}$$

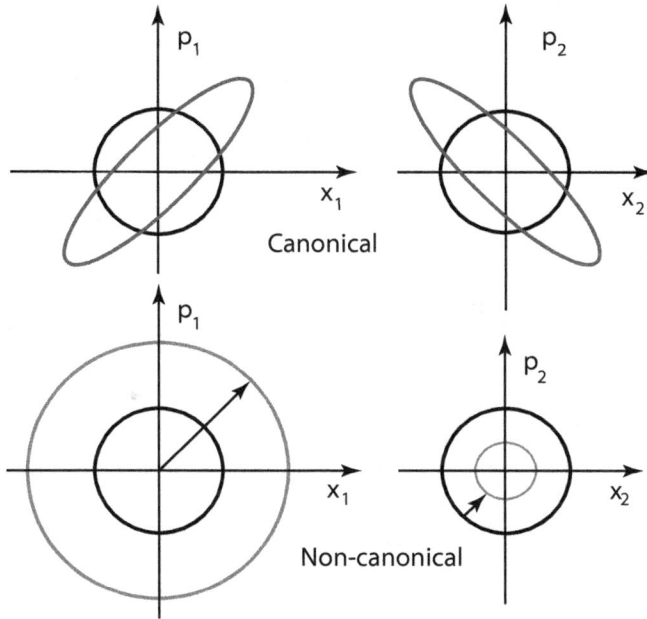

Figure 7.3. Canonical and non-canonical transformations in the x_+, p_+ and x_-, p_- variables. This figure is essentially figure 7.2 rotated by $45°$. Here also, the phase space area is conserved for each mode under the canonical transformation, while this is not the case for non-canonical transformations.

These transformations are illustrated in figure 7.3. Under the transformation of equation (7.86), the phase spaces squeeze in the opposite directions, while preserving their respective areas. This is therefore a canonical transformation. This canonical transformation is generated by Q_1, given in table 7.3.

On the other hand, under the transformation of equation (7.87), the area of one phase space expands while the other area shrinks as also illustrated in figure 7.2 and 7.3. This non-canonical transformation is generated by G_2, also given in table 7.3.

We can rewrite equation (7.86) as

$$
\begin{pmatrix} x'_+ \\ x'_- \\ p'_+ \\ p'_- \end{pmatrix} = \begin{pmatrix} e^{\eta} & 0 & 0 & 0 \\ 0 & e^{-\eta} & 0 & 0 \\ 0 & 0 & e^{-\eta} & 0 \\ 0 & 0 & 0 & e^{\eta} \end{pmatrix} \begin{pmatrix} x_+ \\ x_- \\ p_+ \\ p_- \end{pmatrix}
\tag{7.88}
$$

and equation (7.87) as

$$
\begin{pmatrix} x'_+ \\ x'_- \\ p'_+ \\ p'_- \end{pmatrix} = \begin{pmatrix} e^{\eta} & 0 & 0 & 0 \\ 0 & e^{-\eta} & 0 & 0 \\ 0 & 0 & e^{\eta} & 0 \\ 0 & 0 & 0 & e^{-\eta} \end{pmatrix} \begin{pmatrix} x_+ \\ x_- \\ p_+ \\ p_- \end{pmatrix}.
\tag{7.89}
$$

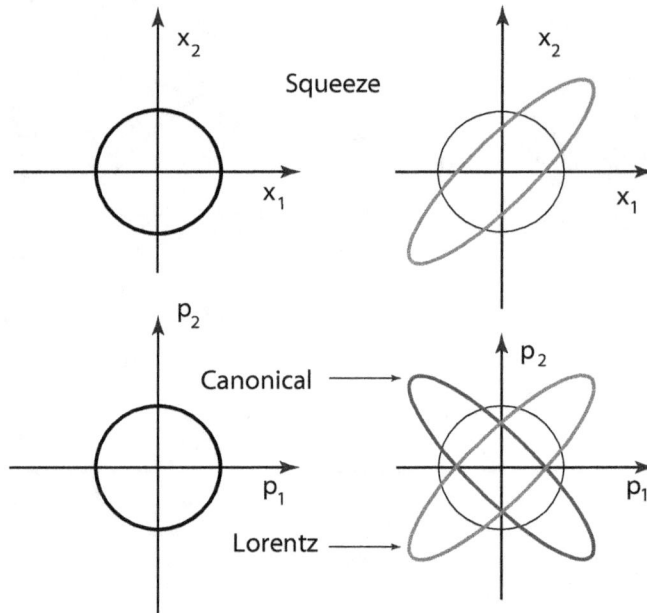

Figure 7.4. Canonical and non-canonical transformations in the x_1, x_2 and p_1, p_2 variables. The horizontal axes are the first mode, and the vertical axes are the second mode. What happens to the horizontal axes corresponding to the first mode when the vertical axes are integrated out as Feynman's rest of the Universe?

These squeeze properties are illustrated in figure 7.2. In Einstein's Lorentz boost, the space–time and momentum–energy coordinates are squeezed in the same direction. In the canonical transformation, they are squeezed in the opposite directions, as illustrated in figure 7.2.

It is possible to give physical interpretations to the expanding phase space, but it is not possible for the shrinking phase space. This shrinking phase space could belong to Feynman's rest of the Universe.

Next, we are interested in the squeeze properties of the (x_1, x_2) and (p_1, p_2) spaces. The squeeze properties are illustrated in figure 7.4. Under the canonical transformation these two spaces are squeezed in the opposite direction. They are, however, squeezed in the same direction under the Lorentz or non-canonical transformation. It is an interesting problem to see what happens to the horizontal axes when the vertical axes are not observed. We shall study this problem in section 8.4.

References

[1] Abraham R and Marsden J E 2008 *Foundations of Mechanics* 2nd edn (Providence, RI: American Mathematical Society)
[2] Başkal S, Kim Y S, and Noz M E 2016 Entangled harmonic oscillators and space–time entanglement *Symmetry* **8** 55–80
[3] Başkal S, Kim Y S, and Noz M E 2019 Poincaré symmetry from Heisenberg's Uncertainty Relations *Symmetry* **11** 409

[4] Davies R W and Davies K T R 1975 On the Wigner distribution function for an oscillator *Ann. Phys.* **89** 261–73

[5] Dirac P A M 1963 A remarkable representation of the 3 + 2 de Sitter group *J. Math. Phys.* **4** 901–9

[6] Feynman R P, Kislinger M, and Ravndal F 1971 Current matrix elements from a relativistic quark model *Phys. Rev.* D **3** 2706–32

[7] Goldstein H 1980 *Classical Mechanics* (*Addison-Wesley Series in Physics*) 2nd edn (Reading, MA: Addison-Wesley)

[8] Han D, Kim Y S, and Noz M E 1995 O (3,3)–like symmetries of coupled harmonic oscillators *J. Math. Phys.* **36** 3940–54

[9] Itzykson C and Zuber J B 2005 *Quantum Field Theory* (*Dover Books on Physics*) (Mineola, NY: Dover)

[10] Kim Y S and Noz M E 1991 *Phase Space Picture of Quantum Mechanics: Group Theoretical Approach* (*Lecture Notes in Physics* vol 40) (Singapore: World Scientific)

[11] Kim Y S and Noz M E 2012 Dirac matrices and Feynman rest of the Universe *Symmetry* **4** 626–43

[12] Kim Y S, Noz M E, and Oh S H 1979 A simple method for illustrating the difference between the homogeneous and inhomogeneous Lorentz groups *Am. J. Phys.* **47** 892–7

[13] Lee D-G 1995 The Dirac gamma matrices as relics of a hidden symmetry? As fundamental representations of the algebra $Sp(4,R)$ *J. Math. Phys.* **36** 524–30

[14] Magnus W, Oberhettinger F, and Soni R P 1966 *Formulas and Theorems for the Special Functions of Mathematical Physics* (Berlin: Springer)

[15] Majorana E 1932 Relativistic particles with arbitrary intrinsic angular momentum *Nuovo Cimento* **9** 335–44

[16] Ruiz M J 1974 Orthogonality relation for covariant harmonic-oscillator wave functions *Phys. Rev.* D **10** 4306–7

[17] Saleh B E A and Carl Teich M 2007 *Fundamentals of Photonics* (*Wiley Series in Pure and Applied Optics*) 2nd edn (Hoboken, NJ: Wiley)

[18] Von Neumann J, Beyer R T, and Wheeler N A 2018 *Mathematical Foundations of Quantum Mechanics* (Princeton, NJ: Princeton University Press)

[19] Walls D F and Milburn G J 2008 *Quantum Optics* 2nd edn (Berlin: Springer)

[20] Wigner E P 1932 On the quantum correction for thermodynamic equilibrium *Phys. Rev.* **40** 749–59

[21] Yuen H P 1976 Two-photon coherent states of the radiation field *Phys. Rev.* A **13** 2226–43

[22] Yurke B, McCall S L, and Klauder J R 1986 $SU(2)$ and $SU(1,1)$ interferometers *Phys. Rev.* A **33** 4033–54

IOP Publishing

Mathematical Devices for Optical Sciences

Sibel Başkal, Young S Kim, and Marilyn E Noz

Chapter 8

Entanglement and entropy

The entanglement problem is one of the fundamental issues in quantum mechanics [2, 3, 17, 23, 32, 36, 41]. Let us consider two dynamical variables. If one variable is completely independent of what happens to the other variable, we can write the wave function as a direct product of two wave functions. However, if the wave function is not separable, we face the entanglement problem. Most often, we face the problem of what happens to the first dynamical variable when the second variable is not observed.

For this case, in his book on statistical mechanics [15], Feynman made the following statement:

> When we solve a quantum-mechanical problem, what we really do is divide the Universe into two parts—the system in which we are interested and the rest of the Universe. We then usually act as if the system in which we are interested comprised the entire Universe. To motivate the use of density matrices, let us see what happens when we include the part of the Universe outside the system.

The best way to illustrate this concept of Feynman's rest of the Universe is to use two coupled oscillators, where one of them belongs to the part of the Universe in which we do physics while the second oscillator belongs to the rest of the Universe [22, 28]. The question is what happens to the first oscillator when the second oscillator is not observed.

For dealing with this problem, there is a mathematical device called the density matrix. The density matrix was introduced in chapter 1 as the basic mathematical device to calculate measurable quantities from wave functions. In this chapter, we study this matrix in order to understand entanglement problems.

Among the entanglement problems, two systems coupled by a two-by-two matrix or by a Gaussian function are most often discussed in the physics literature. Indeed, Gaussian entanglement is exemplified by the squeezed states of light discussed in chapter 7. Additionally, Gaussian entanglement is of current interest, and is

doi:10.1088/2053-2563/aafe78ch8   8-1

discussed extensively in the literature [6, 18, 19, 27]. It is also of interest in other dynamical systems [1, 3, 4, 9, 13, 20, 27]. The underlying mathematical language for this form of entanglement is that of harmonic oscillators.

In section 8.1 we define first the density matrix and discuss how it is constructed. The density matrix stands between the information contained in a given quantum system and the quantities we can measure. If not all the quantities are measurable, they appear as a statistical quantity known as the entropy of the system. In section 8.2 we use two-by-two matrices to spell out the basic ideas of the density matrix.

In section 8.3 the density matrix is constructed for the system of two coupled oscillators, and thus for two-mode coherent states in quantum optics and for the Gaussian entanglements. In section 8.4 we discuss in detail how the entropy is generated from this two-mode system. In section 8.5 the density matrix formalism is extended to harmonic oscillators for excited states. The formulae are given for the density matrix and the entropy.

Finally, in section 8.6, we study the entanglement problem using Wigner functions. It is shown that the failure to measure the second variable results in the addition of statistical uncertainty, thus increasing the areas of total uncertainty in phase space.

8.1 Density matrix and entropy

We could start with a wave function depending on $x_1, x_2, \ldots, x_n$ variables, but it is sufficient to study a wave function with two variables, namely $\psi(x_1, x_2)$. Then the density matrix is defined as

$$\rho(x_1, x_2; x_1', x_2') = \psi(x_1, x_2)\psi^*(x_1', x_2'). \tag{8.1}$$

Its trace is

$$\mathrm{Tr}(\rho) = \int \rho(x_1, x_2; x_1, x_2) \, dx_1 \, dx_2. \tag{8.2}$$

The result of this integration is 1, and thus $\mathrm{Tr}(\rho) = 1$. We can also compute $\mathrm{Tr}(\rho^2)$ as

$$\mathrm{Tr}(\rho^2) = \int \left[\int \rho(x_1, x_2; x_1', x_2')\rho(x_1', x_2'; x_1, x_2) \, dx_1' \, dx_2' \right] dx_1 \, dx_2$$
$$= \int \rho(x_1, x_2; x_1, x_2) \, dx_1 \, dx_2. \tag{8.3}$$

This means $\mathrm{Tr}(\rho^2) = 1$. Here we have not done anything on either coordinate. In this case, the density matrix defined in equation (8.1) is called the pure state.

If the second variable x_2 is not observed, we have to integrate over this variable, and the resulting density matrix depends only the first variable. The density matrix becomes

$$\rho_1(x, x') = \int \rho(x, x_2; x', x_2) \, dx_2, \tag{8.4}$$

where x_1 and x_1' have been replaced by x and x', respectively. The trace of this density matrix is

$$\mathrm{Tr}(\rho) = \int \rho(x, x)\, dx, \tag{8.5}$$

which becomes the same as equation (8.2). On the other hand for $\mathrm{Tr}(\rho^2)$, the story is different:

$$\mathrm{Tr}(\rho^2) = \int \rho_1(x, x')\rho(x', x)\, dx'\, dx. \tag{8.6}$$

This expression is different from that of $\mathrm{Tr}(\rho^2)$ in equation (8.3), and will be less than 1 [5, 12, 15, 31]. We shall discuss these issues in sections 8.2 and 8.3 using a two-by-two density matrix and two-mode squeezed states, respectively [3].

We can study the density matrix in terms of the orthonormal expansion of the wave function, such as

$$\psi(x_1, x_2) = \sum_{n_1 n_2} A_{n_1, n_2}\, \chi_{n_1}(x_1)\chi_{n_2}(x_2), \tag{8.7}$$

with

$$\int \chi_n^*(x)\chi_m(x)\, dx = \delta_{nm} \quad \text{and} \quad \sum_{mn} A_{nm}^* A_{nm} = 1. \tag{8.8}$$

Thus the pure-state density matrix should be

$$\rho(x_1, x_2; x_1', x_2') = \sum_{n_1 n_2} \sum_{m_1 m_2} A_{m_1 m_2} A_{n_1 n_2}^*\, \chi_{m_1}(x_1)\chi_{m_2}(x_2)\chi_{n_1}^*(x_1')\chi_{n_2}^*(x_2'). \tag{8.9}$$

If x_2 is not measured, and if we use x for x_1, the density matrix $\rho(x, x')$ takes the form

$$\rho(x, x') = \sum_{m,n} \rho_{mn}\, \chi_m(x)\chi_n^*(x'), \tag{8.10}$$

with

$$\rho_{mn}(\alpha, \eta) = \sum_{k} A_{mk}(\alpha, \eta)A_{nk}^*(\alpha, \eta), \tag{8.11}$$

after integrating over the x_2 variable. The matrix $\rho_{mn}(\alpha, \eta)$ is also called the density matrix. This matrix is Hermitian and can therefore be diagonalized.

If the diagonal elements are ρ_{kk}, the entropy of the system is defined as [40, 43]

$$S = -\sum_{k} \rho_{kk} \ln(\rho_{kk}). \tag{8.12}$$

The entropy is zero for a pure state, and increases as the system becomes impure. Like $\mathrm{Tr}(\rho^2)$, this quantity is a measure of our ignorance of the coordinate x_2.

8.2 Two-by-two density matrices

In order to gain insight about the concept of entanglement, let us consider two spin states:

$$|0\rangle = \begin{pmatrix} 1 \\ 0 \end{pmatrix} e^{-i\omega_1 t} \quad \text{and} \quad |1\rangle = \begin{pmatrix} 0 \\ 1 \end{pmatrix} e^{-i\omega_2 t}. \tag{8.13}$$

If the first spin state $|0\rangle$ depends on what happens to the second state $|1\rangle$, we say the system is entangled.

Let us next consider the linear combination

$$|\alpha\rangle = \cos\alpha|0\rangle + \sin\alpha|1\rangle = \begin{pmatrix} (\cos\alpha)e^{-i\omega_1 t} \\ (\sin\alpha)e^{-i\omega_2 t} \end{pmatrix}. \tag{8.14}$$

In this case, what happens to the state $|0\rangle$ depends on what happens to the state $|1\rangle$. Most often, we face the problem of what happens when the state $|1\rangle$ is not observed. This is what entanglement is all about.

In his book on statistical mechanics [15], Feynman illustrates the problem using two-by-two matrices. Let us expand what Feynman did for the two-state problem. According to equation (8.1), the density matrix for the two states is defined as

$$\rho(\alpha) = |\alpha\rangle\langle\alpha| = \begin{pmatrix} \cos^2\alpha & \cos\alpha\sin\alpha\,e^{-i\delta t} \\ \cos\alpha\sin\alpha\,e^{i\delta t} & \sin^2\alpha \end{pmatrix}, \tag{8.15}$$

with $\delta = \omega_1 - \omega_2$. This density matrix satisfies the pure-state conditions: $[\rho(\alpha)]^2 = \rho(\alpha)$ and $\mathrm{Tr}(\rho) = \mathrm{Tr}(\rho^2) = 1$. These are the properties of pure states where we have not done anything to the system [35].

Among the entanglement problems, we most often encounter the case where the second spinor is not observed. In this case, we have to take the arithmetic average of the probabilities of the second spinor, namely the cases with $\alpha = 0$ and $\alpha = 90°$. The density matrix becomes

$$\rho = \begin{pmatrix} 1/2 & 0 \\ 0 & 1/2 \end{pmatrix}. \tag{8.16}$$

Here $\mathrm{Tr}(\rho)$ is still one. However, $\mathrm{Tr}(\rho^2) < 1$. This is the effect of our failure to be able to make measurements on the state $|1\rangle$.

One way to measure this degree of our ignorance is to calculate the entropy of the system. The density matrix is Hermitian, and thus can be diagonalized. Let the diagonal elements be ρ_{11} and ρ_{22}. Then the entropy of the system is

$$S = -\sum_k \rho_{kk} \ln(\rho_{kk}). \tag{8.17}$$

Let us go back to the pure state given in equation (8.15). This two-by-two matrix is Hermitian, and thus can be diagonalized. The diagonal elements are 0 and 1. Thus the entropy of this pure state is zero.

Let us next go to the diagonal matrix of equation (8.16) for the state where the second state is not measured. The entropy in this case is

$$S = -\frac{1}{2} \ln \left(\frac{1}{2}\right) - \frac{1}{2} \ln \left(\frac{1}{2}\right) = \ln(2). \tag{8.18}$$

If $\delta \neq 0$, the time average of the density matrix in equation (8.15) is

$$\bar{\rho} = \begin{pmatrix} \cos^2 \alpha & 0 \\ 0 & \sin^2 \alpha \end{pmatrix}. \tag{8.19}$$

The entropy derivable from this density matrix is

$$S = -(\cos^2 \alpha) \ln(\cos^2 \alpha) - (\sin^2 \alpha) \ln(\sin^2 \alpha). \tag{8.20}$$

Here also, $S = 0$ when $\alpha = 0$ or $90°$. It becomes $\ln(2)$ when $\alpha = 45°$.

8.3 Density matrix for two-oscillator states

Let us start with the ground-state wave function of two coordinate variables which takes the form

$$\psi_{0,0}(x_1, x_2) = \frac{1}{\sqrt{\pi}} \exp \left\{ -\frac{1}{2}(x_1^2 + x_2^2) \right\}. \tag{8.21}$$

This wave function corresponds to the vacuum state of two different photons. This function is separable in the x_1 and x_2 coordinates, which are thus not entangled.

As was seen in chapter 7, the coordinates can be rotated and squeezed. Under the coordinate rotation

$$\begin{pmatrix} y_1 \\ y_2 \end{pmatrix} = \begin{pmatrix} \cos \alpha & -\sin \alpha \\ \sin \alpha & \cos \alpha \end{pmatrix} \begin{pmatrix} x_1 \\ x_2 \end{pmatrix}, \tag{8.22}$$

the wave function remains separable even though it is now written as

$$\psi_{0,0}(x_1, x_2) = \frac{1}{\sqrt{\pi}} \exp \left\{ -\frac{1}{2}\left[(x_1 \cos \alpha + x_2 \sin \alpha)^2 \right.\right.$$
$$\left.\left. + (-x_1 \sin \alpha + x_2 \cos \alpha)^2 \right] \right\}. \tag{8.23}$$

We can next consider the squeeze transformation:

$$\begin{pmatrix} z_1 \\ z_2 \end{pmatrix} = \begin{pmatrix} e^\eta & 0 \\ 0 & e^{-\eta} \end{pmatrix} \begin{pmatrix} y_1 \\ y_2 \end{pmatrix}. \tag{8.24}$$

The wave function then becomes

$$\psi_{\alpha,\eta}(x_1, x_2) = \frac{1}{\sqrt{\pi}} \exp \left\{ -\frac{1}{2}\left[e^{-2\eta}(x_1 \cos \alpha + x_2 \sin \alpha)^2 \right.\right.$$
$$\left.\left. + e^{2\eta}(-x_1 \sin \alpha + x_2 \cos \alpha)^2 \right] \right\}. \tag{8.25}$$

This wave function remains separatable in the x_1 and x_2 variables unless both the α and η variables are non-zero, as indicated in table 8.1.

According to the definition given in equation (8.1), the density matrix is

$$\rho_{\alpha,\eta}(x_1, x_2; x_1', x_2') = \psi_{\alpha,\eta}(x_1, x_2)\psi_{\alpha,\eta}^*(x_1', x_2'),\tag{8.26}$$

which can be written as

$$
\begin{aligned}
\rho_{\alpha,\eta} = \frac{1}{\pi}\exp\Bigg\{ &-\frac{1}{2}e^{-2\eta}\Big[(x_1\cos\alpha + x_2\sin\alpha)^2 \\
&+ (x_1'\cos\alpha + x_2'\sin\alpha)^2\Big]\Bigg\} \\
\times\exp\Bigg\{ &-\frac{1}{2}e^{2\eta}\Big[(-x_1\sin\alpha + x_2\cos\alpha)^2 \\
&+ (-x_1'\sin\alpha + x_2'\cos\alpha)^2\Big]\Bigg\}.
\end{aligned}
\tag{8.27}
$$

This expression is consistent with the condition

$$\mathrm{Tr}\,(\rho_{\alpha,\eta}) = \mathrm{Tr}\left(\rho_{\alpha,\eta}^2\right) = 1\tag{8.28}$$

for pure states.

If no observations are made on the x_2 variable, we can follow the procedure given in equation (8.4). The resulting density matrix for x_1 is

$$\rho(x, x') = \int \rho(x, x_2; x', x_2)\, dx_2,\tag{8.29}$$

where x_1 is replaced with x. The evaluation of this integral leads to

$$
\rho(x, x') = \left[\frac{1}{\pi(\cosh(2\eta) - \sinh(2\eta)\cos(2\alpha))}\right]^{1/2}
$$
$$
\times\exp\left\{-\left[\frac{(x + x')^2 + (x - x')^2(\cosh^2(2\eta) - \sinh^2(2\eta)\cos^2(2\alpha))}{4(\cosh(2\eta) - \sinh(2\eta)\cos(2\alpha))}\right]\right\}.
\tag{8.30}
$$

With this expression, we can check the trace relations for the density matrix. Indeed the trace integral

Table 8.1. Two entangled oscillators. Unless both parameters are non-zero, the variables x_1 and x_2 remain separated.

	$\alpha = 0$	$\alpha \neq 0$
$\eta = 0$	Separated	Separated
$\eta \neq 0$	Separated	Entangled

$$\text{Tr}(\rho) = \int \rho(x, x)\, dx \tag{8.31}$$

becomes 1, as in the case of all density matrices. As for $\text{Tr}(\rho^2)$, the result of the trace integral becomes

$$\text{Tr}(\rho^2) = \int \rho(x, x')\rho(x', x)\, dx'dx = \frac{1}{\sqrt{1 + \sinh^2(2\eta)\sin^2(2\alpha)}}. \tag{8.32}$$

This is less than 1 for non-zero values of η. This is consistent with the general theory of density matrices. If $\alpha = 0$ and/or $\eta = 0$, the first oscillator is totally independent of the second oscillator, and the system of the first oscillator is in a pure state, and $\text{Tr}(\rho^2)$ becomes 1. This result is also consistent with the general theory of density matrices.

8.4 Entropy for the two-mode state

Among the rotation angles, $\alpha = 45°$ is the most interesting angle to study. It is the normal coordinate system. The two-mode squeezed state is precisely for this angle. As given in chapter 7, the wave function is

$$\psi(x_1, x_2) = \frac{1}{\sqrt{\pi}} \exp\left\{-\frac{1}{4}\left[e^{-2\eta}(x_1 + x_2)^2 + e^{2\eta}(x_1 - x_2)^2\right]\right\}. \tag{8.33}$$

This wave function can be expanded as

$$\psi(x_1, x_2) = \frac{1}{\cosh\eta} \sum_k (\tanh\eta)^{2k}\, \chi_k(x_1)\chi_k(x_2), \tag{8.34}$$

where $\chi_k(x)$ is the kth excited-state oscillator wave function. The density matrix becomes

$$\rho(x_1, x_2; x_1', x_2') = \frac{1}{\cosh^2\eta} \sum_{k,k'} (\tanh\eta)^k(\tanh\eta)^{k'}$$
$$\times \chi_k(x_1)\chi_k(x_2)\chi_{k'}^*(x_1)\chi_{k'}^*(x_2). \tag{8.35}$$

The trace of this matrix is

$$\frac{1}{\cosh^2\eta} \sum_k (\tanh\eta)^{2k} = 1. \tag{8.36}$$

If the x_2 variable is not observed,

$$\rho(x, x') = \left(\frac{1}{\cosh(\eta)}\right)^2 \sum_k (\tanh\eta)^{2k}\, \chi_k(x)\chi_k^*(x'), \tag{8.37}$$

which leads to $\text{Tr}(\rho) = 1$. It is also straightforward to compute the integral for $\text{Tr}(\rho^2)$. The calculation leads to

$$\mathrm{Tr}(\rho^2) = \left(\frac{1}{\cosh\eta}\right)^4 \sum_k (\tanh\eta)^{4k} = \left(\frac{1}{\cosh\eta}\right)^4 \frac{1}{1-\tanh^4\eta}. \tag{8.38}$$

Since

$$1 - \tanh^4\eta = (1 - \tanh^2\eta)(1 + \tanh^2\eta) = \cosh(2\eta)/\cosh^4\eta, \tag{8.39}$$

$$\mathrm{Tr}(\rho^2) = \frac{1}{\cosh(2\eta)} < 1. \tag{8.40}$$

This result is consistent with equation (8.32) for $\alpha = 45°$. This is of course due to the fact that x_2 is not observed and therefore we have averaged over it.

However, the standard way to measure this ignorance is to calculate the entropy defined as [15, 40, 43]

$$S = -\mathrm{Tr}\,(\rho\,\ln(\rho)). \tag{8.41}$$

Indeed, the ρ matrix given in equation (8.37) is already diagonalized. The diagonal elements are

$$\rho_{kk} = \frac{(\tanh\eta)^{2k}}{\cosh^2\eta}. \tag{8.42}$$

Thus

$$\rho_{kk}\ln(\rho_{kk}) = \frac{(\tanh\eta)^{2k}}{\cosh^2\eta}[-\ln(\cosh^2\eta) + k\,\ln(\tanh^2\eta)] \tag{8.43}$$

and

$$\sum_k \rho_{kk}\ln(\rho_{kk}) = \frac{1}{\cosh^2\eta}\sum_k\Big\{-\ln(\cosh^2\eta)\tanh^{2k}\eta$$
$$+ \ln(\tanh^2\eta)k\,\tanh^{2k}\eta\Big\}. \tag{8.44}$$

There are two summations in this expression, namely

$$\sum_k(\tanh^2\eta)^k = \cosh^2\eta \quad\text{and}\quad \sum_k k(\tanh^2\eta)^k = (\sinh^2\eta)\cosh^2\eta, \tag{8.45}$$

therefore

$$\sum_k \rho_{kk}\ln(\rho_{kk}) = -\ln(\cosh^2\eta) + (\sinh^2\eta)[\ln(\sinh^2\eta) - \ln(\cosh^2\eta)]. \tag{8.46}$$

The entropy then becomes

$$S = (\cosh^2\eta)\ln(\cosh^2\eta) - (\sinh^2\eta)\ln(\sinh^2\eta). \tag{8.47}$$

This expression can be translated into a more familiar form if we use the notation

$$\tanh^2(\eta) = \exp\left(\frac{-\hbar\omega}{k_B T}\right), \tag{8.48}$$

where $\hbar\omega$ is the energy separation of the oscillator system, k_B is Boltzmann's constant, and T is the absolute temperature. The ratio $\hbar\omega/k_B T$ is a dimensionless variable. In terms of this variable, this entropy takes the form [21, 30]

$$S = \left(\frac{\hbar\omega}{k_B T}\right)\frac{1}{\exp(\hbar\omega/k_B T) - 1} - \ln\left[1 - \exp(-\hbar\omega/k_B T)\right]. \tag{8.49}$$

This familiar expression is for the entropy of an oscillator state in thermal equilibrium. Thus, for this oscillator system, we can relate our ignorance to the temperature. It is interesting to note that the squeeze parameter η can be related to the temperature variable.

8.5 Entangled excited states

In section 8.4 we studied the entanglement problem when the two-dimensional Gaussian function is squeezed. It is thus of interest to see what happens when the two-dimensional harmonic oscillator is excited. In this problem, the oscillator can be excited in both variables, and this case was studied by Rotbart in 1981 [37].

In this section, let us study the case where there are no excitations for the x_2 variable, as in the case of the c-number time–energy uncertainty relation [8, 30], or as in the case of the shadow coordinate in thermo-field dynamics [14, 34, 39].

For this system, we can start with the wave function

$$\psi(x_1, x_2) = \chi_n(x_1)\chi_0(x_2)$$
$$= \left[\frac{1}{\pi\, 2^n n!}\right]^{1/2} H_n(x_1)\exp\left\{-\frac{1}{2}(x_1^2 + x_2^2)\right\}. \tag{8.50}$$

After the squeeze transformation, we should replace x_1 and x_2 by y_1 and y_2, respectively, where

$$\begin{pmatrix} y_1 \\ y_2 \end{pmatrix} = \begin{pmatrix} \cosh\eta & -\sinh\eta \\ -\sinh\eta & \cosh\eta \end{pmatrix}\begin{pmatrix} x_1 \\ x_2 \end{pmatrix}. \tag{8.51}$$

The wave function becomes

$$\psi_\eta(y_1, y_2) = \chi_n(y_1)\chi_0(y_2)$$
$$= \left[\frac{1}{\pi\, 2^n n!}\right]^{1/2} H_n(y_1)\exp\left\{-\frac{1}{2}(y_1^2 + y_2^2)\right\}. \tag{8.52}$$

This form could be written as

$$\psi_\eta(x_1, x_2) = \sum_{k',k} A_{n,m}\, \chi_{k'}(x_1)\chi_k(x_2), \tag{8.53}$$

and each term in this series should satisfy the differential equation [16]

$$\frac{1}{2}\left\{\left(x_1^2 - \frac{\partial^2}{\partial x_1^2}\right) - \left(x_2^2 - \frac{\partial^2}{\partial x_2^2}\right)\right\}\chi_n(x_1)\chi_m(x_2) = (n - m)\chi_n(x_1)\chi_m(x_2). \qquad (8.54)$$

This means that $(k' - k)$ is a constant number. Thus, we can let $k' = k + n$. We can write the series as a single sum:

$$\psi_\eta(x_1, x_2) = \sum_k A_k(n)\chi_{(k+n)}(x_1)\chi_k(x_2), \qquad (8.55)$$

with

$$\sum_k |A_k(n)|^2 = 1. \qquad (8.56)$$

This coefficient is

$$A_k(n) = \int \chi_{k+n}(x_1)\chi_k(x_2)\psi_\eta(x_1, x_2)\, dx_1\, dx_2. \qquad (8.57)$$

This calculation was given in the literature in a fragmentary way in connection with a Lorentz-covariant description of extended particles starting from Ruiz's 1974 paper [38], subsequently by Kim *et al* in 1979 [29] and by Rotbart in 1981 [37]. In view of the recent developments in physics [4], it seems necessary to give one coherent calculation of the coefficient of equation (8.53).

We are now interested in the squeezed oscillator function

$$A_k(n) = \left[\frac{1}{\pi^2\, 2^n n!(k + n)^2(n + k)!k^2 k!}\right]^{1/2}$$

$$\times \int H_{n+k}(x_1)H_k(x_2)H_n(y_1) \exp\left\{-\left(\frac{x_1^2 + x_2^2 + y_1^2 + y_2^2}{2}\right)\right\} dx_1 dx_2\, dy_1 dy_2. \qquad (8.58)$$

As was noted by Ruiz [38], the key to the evaluation of this integral is to introduce the generating function for the Hermite polynomials [10, 33],

$$G(r, z) = \exp(-r^2 + 2rz) = \sum_m \frac{r^m}{m!}H_m(z), \qquad (8.59)$$

and evaluate the integral

$$I = \int G(r, x_1)G(s, x_2)G(r', y_1)\exp\left\{-\left(\frac{x_1^2 + x_2^2 + y_1^2 + y_2^2}{2}\right)\right\} dx_1 dx_2\, dy_1 dy_2. \qquad (8.60)$$

The integrand becomes one exponential function, and its exponent is quadratic in the x and y variables. This quadratic form can be diagonalized, and the integral can be evaluated [25, 29]. The result is

$$I = \left[\frac{\pi}{\cosh \eta}\right] \exp(2rs \tanh \eta) \exp\left(\frac{2rr'}{\cosh \eta}\right). \tag{8.61}$$

We can now expand this expression and choose the coefficients of r^{n+k}, s^k, r'^n for $H_{(n+k)}(x_1)$, $H_n(x_2)$, and $H_n(y_1)$, respectively. The result is

$$A_k(n) = \left(\frac{1}{\cosh \eta}\right)^{(n+1)} \left[\frac{(n+k)!}{n!k!}\right]^{1/2} (\tanh \eta)^k. \tag{8.62}$$

Thus, the series becomes

$$\psi(x_1, x_2) = \left(\frac{1}{\cosh \eta}\right)^{(n+1)} \sum_k \left[\frac{(n+k)!}{n!k!}\right]^{1/2} (\tanh \eta)^k \chi_{k+n}(x_1)\chi_k(x_2). \tag{8.63}$$

If $n = 0$, it is the squeezed ground state, and this expression becomes the entangled state of the Gaussian function discussed in section 8.4.

From this wave function, we can construct the pure-state density matrix

$$\rho_\eta^n(x_1, x_2; x_1', x_2') = \psi_\eta^n(x_1, x_2)\psi_\eta^{n*}(x_1', x_2'), \tag{8.64}$$

which satisfies the condition $\rho^2 = \rho$:

$$\rho_\eta^n(x_1, x_2; x_1', x_2') = \int \rho_\eta^n(x_1, x_2; z_1, z_2)\rho_\eta^n(z_1, z_2; x_1', x_2') \, dz_1 \, dz_2. \tag{8.65}$$

If we are not able to measure the x_2 variable, the density matrix becomes

$$\begin{aligned}
\rho_\eta^n(x, x') &= \int \psi_\eta^n(x, x_2)\psi_\eta^{n*}(x', x_2) \, dx_2 \\
&= \left(\frac{1}{\cosh \eta}\right)^{2(n+1)} \sum_k \frac{(n+k)!}{n!k!} \\
&\quad \times (\tanh \eta)^{2k}\psi_{n+k}(x)\psi_{k+n}^*(x'),
\end{aligned} \tag{8.66}$$

where x_1 and x_1' are replaced by x and x', respectively.

The trace of this density matrix is 1, but the trace of ρ^2 is less than 1, as

$$\begin{aligned}
\mathrm{Tr}\,(\rho^2) &= \int \rho_\eta^n(x, x')\rho_\eta^n(x, x')dxdx' \\
&= \left(\frac{1}{\cosh \eta}\right)^{4(n+1)} \sum_k \left[\frac{(n+k)!}{n!k!}\right]^2 (\tanh \eta)^{4k},
\end{aligned} \tag{8.67}$$

which is less than 1.

The standard way to measure this ignorance is to calculate the entropy defined as [40, 43]

$$S = -\mathrm{Tr}\,(\rho \ln(\rho)). \tag{8.68}$$

If we pretend to know the distribution along the x_2-direction, the entropy is zero. However, if we do not know how to deal with distribution along the x_2-direction, we should use the density matrix of equation (8.66) to calculate the entropy as we did in section 8.4, and the result is

$$S = 2(n + 1)[(\cosh^2 \eta) \ln(\cosh^2 \eta) - (\sinh^2 \eta) \ln(\sinh^2 \eta)]$$

$$- \left(\frac{1}{\cosh \eta}\right)^{2(n+1)} \sum_k \frac{(n + k)!}{n!k!} \ln\left[\frac{(n + k)!}{n!k!}\right](\tanh \eta)^{2k}. \qquad (8.69)$$

This expression becomes the entropy of equation (8.47) when $n = 0$.

8.6 Wigner functions and uncertainty relations

As was discussed in chapter 5, the Wigner function is derived from the density matrix according to [26, 42]

$$W(x_1, x_2; p_1, p_2) = \left(\frac{1}{\pi}\right)^2 \int \exp\{-2i(p_1 x_1' + p_2 x_2')\}$$

$$\times \psi^*(x_1 + x_1', x_2 + x_2')\psi(x_1 - x_1', x_2 - x_2') \, dx_1' \, dx_2'. \qquad (8.70)$$

If the wave function $\psi(x_1, x_2)$ is of the Gaussian form for the ground state, the Wigner function becomes

$$W(x_1, x_2; p_1, p_2) = \left(\frac{1}{\pi}\right)^2 \exp\left\{-\frac{1}{2}(x_1^2 + x_2^2 + p_1^2 + p_2^2)\right\}. \qquad (8.71)$$

This Wigner function is normalized as

$$\int W(x_1, x_2; p_1, p_2) \, dx_1 \, dp_1 \, dx_2 \, dp_2 = 1. \qquad (8.72)$$

As is given in equation (8.25), the Gaussian wave function can be squeezed to

$$\psi_{\alpha,\eta}(x_1, x_2) = \frac{1}{\sqrt{\pi}} \exp\left\{-\frac{1}{2}\left[e^{2\eta}(-x_1 \sin \alpha + x_2 \cos \alpha)^2\right.\right.$$

$$\left.\left. + e^{-2\eta}(x_1 \cos \alpha + x_2 \sin \alpha)^2\right]\right\}. \qquad (8.73)$$

This is an entangled wave function. It is possible to construct a density matrix and then the entangled Wigner function of the form

$$W_{\alpha,\eta}(x_1, x_2; p_1, p_2) = \left(\frac{1}{\pi}\right)^2 \exp\{-e^{-2\eta}(x_1 \cos \alpha + x_2 \sin \alpha)^2$$

$$- e^{2\eta}(-x_1 \sin \alpha + x_2 \cos \alpha)^2$$

$$- e^{-2\eta}(p_1 \cos \alpha + p_2 \sin \alpha)^2$$

$$- e^{2\eta}(-p_1 \sin \alpha + p_2 \cos \alpha)^2\}. \qquad (8.74)$$

The key question is what happens in the world of x_1 when the x_2 variable is not observed. We have to integrate the Wigner function over x_2 and p_2, and the result of this integration is [11, 22, 28, 44]

$$W(x_1, p_1) = \int W(x_1, x_2; p_1, p_2) \, dx_2 \, dp_2, \tag{8.75}$$

which becomes

$$W(x, p) = \left[\frac{1}{\pi^2(1 + \sinh^2(2\eta)\sin^2(2\alpha))} \right]^{1/2}$$
$$\times \exp\left\{ -\left[\frac{x^2}{\cosh(2\eta) - \sinh(2\eta)\cos(2\alpha)} \right.\right. \tag{8.76}$$
$$\left.\left. + \frac{p^2}{\cosh(2\eta) + \sinh(2\eta)\cos(2\alpha)} \right]\right\},$$

where we have replaced x_1 and p_1 by x and p, respectively. This Wigner function gives an elliptic distribution in the phase space of x and p. The area of the ellipse is

$$\pi\left[1 + \sin^2(2\alpha)\sinh^2(2\eta) \right]^{1/2}, \tag{8.77}$$

which measures the uncertainty. The minimum uncertainty is π, and the statistical uncertainty coming from our ignorance of the x_2 and p_2 variables is

$$\pi\left[\sin^2(2\alpha)\sinh^2(2\eta) \right]^{1/2}. \tag{8.78}$$

This statistical uncertainty vanishes when the system becomes separated or untangled with $\alpha = 0$, and/or $\eta = 0$. If $\alpha = 45°$, the Wigner function of equation (8.74) becomes

$$W(x, p) = \frac{1}{\pi} \exp\left\{ -\left(\frac{x^2 + p^2}{\cosh(2\eta)} \right) \right\}, \tag{8.79}$$

with a circular distribution in phase space.

When $\alpha = 0$, the system becomes untangled, but the distribution in x and p becomes squeezed. However, since the system is not entangled, there are no effects from the x_2 and p_2 variables. If $\alpha = 0$, as we noted in chapter 7, the Wigner function can be squeezed in the following three ways,

$$W_1 = \left(\frac{1}{\pi} \right)^2 \exp\left\{ e^{-2\eta}x_1^2 + e^{2\eta}p_1^2 + e^{2\eta}x_2^2 + e^{-2\eta}p_2^2 \right\}, \tag{8.80}$$

$$W_2 = \left(\frac{1}{\pi} \right)^2 \exp\left\{ e^{-2\eta}x_1^2 + e^{2\eta}p_1^2 + e^{-2\eta}x_2^2 + e^{2\eta}p_2^2 \right\}, \tag{8.81}$$

$$W_3 = \left(\frac{1}{\pi}\right)^2 \exp\left\{e^{-2\eta}\left(x_1^2 + p_1^2\right) + e^{2\eta}\left(x_2^2 + p_2^2\right)\right\}. \tag{8.82}$$

All these squeezed Wigner functions are separable in the $x_1\, p_1$ and the $x_2\, p_2$ variables. The W_1 and W_2 distributions are area-preserving transformations, and thus are canonical transformations. On the other hand, W_3 becomes

$$W_3 = \frac{1}{\pi} \exp\left\{e^{2\eta}\left(x_1^2 + p_1^2\right)\right\}, \tag{8.83}$$

after the integrations over the second set of variables. Let us go back to the form of equation (8.82). There are two separate phase spaces. If one of the areas increases, the other is decreased. It could go below the uncertainty limit. In this case, these variables are not observable [28].

Let us go back to equation (8.71). The ground-state Wigner function can also be written as

$$W = \left(\frac{1}{\pi}\right)^2 \exp\left\{-\frac{1}{2}\left[(x_1 + x_2)^2 + (x_1 - x_2)^2\right.\right.$$
$$\left.\left. + (p_1 + p_2)^2 + (p_1 - p_2)^2\right]\right\}. \tag{8.84}$$

This Wigner function can also be squeezed in the three different ways, with the quadratic forms

$$Qf_1 = \left[e^{-2\eta}\left\{(x_1 + x_2)^2 + (p_1 + p_2)^2\right\}\right.$$
$$\left. + e^{2\eta}\left\{(x_1 - x_2)^2 + (p_1 - p_2)^2\right\}\right],$$
$$Qf_2 = \left[e^{-2\eta}\left\{(x_1 + x_2)^2 + (p_1 - p_2)^2\right\}\right.$$
$$\left. + e^{2\eta}\left\{(x_1 - x_2)^2 + (p_1 + p_2)^2\right\}\right], \tag{8.85}$$
$$Qf_3 = \left[e^{-2\eta}\left\{(x_1 + x_2)^2 + (x_1 - x_2)^2\right\}\right.$$
$$\left. + e^{2\eta}\left\{(p_1 + p_2)^2 + (p_1 - p_2)^2\right\}\right].$$

Among these three quadratic forms, Qf_3 becomes

$$Qf_3 = 2\left[\left(e^{-2\eta}x_1^2 + e^{2\eta}p_1^2\right) + \left(e^{-2\eta}x_2^2 + e^{2\eta}p_2^2\right)\right]. \tag{8.86}$$

Thus, the squeezed Wigner function with this quadratic form becomes separable, and the problem becomes the same as the Wigner function W_3 of equation (8.82).

The squeezed Wigner functions with the quadratic forms of Qf_1 and Qf_2 are squeezed as described in figure 8.1. For both cases, if the x_2 and p_2 variables are integrated out, the Wigner function becomes

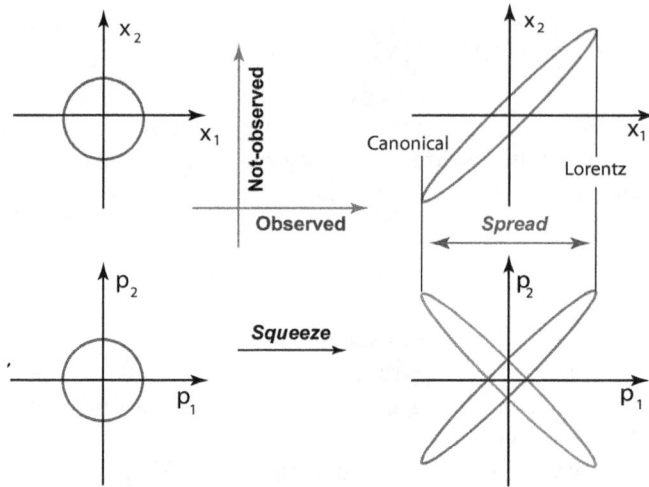

Figure 8.1. Squeeze transformations in the phase space. They could lead to canonical and to Lorentz transformations. In both cases, the widths of space and momentum distribution increase in the same way. This figure is essentially the same as figure 7.4 of chapter 7. However, it is important to note that the variables in the vertical direction are not measurable, while those in the horizontal direction are measurable [4].

$$W(x, p) = \frac{1}{\pi} \exp\left\{-\left(\frac{x_1^2 + p_1^2}{\cosh(2\eta)}\right)\right\}, \tag{8.87}$$

with a circular distribution in phase space [7, 24]. This formula is consistent with the distribution given in equation (8.79).

References

[1] Adesso G and Illuminati F 2007 Entanglement in continuous-variable systems: recent advances and current perspectives *J. Phys. A: Math. Theor.* **40** 7821–80

[2] Barnett S M and Phoenix S J D 1991 Information theory, squeezing, and quantum correlations *Phys. Rev. A* **44** 535–45

[3] Başkal S, Kim Y S, and Noz M E 2016 Entangled harmonic oscillators and space-time entanglement *Symmetry* **8** 55–80

[4] Başkal S, Kim Y S, and Noz M E 2015 *Physics of the Lorentz Group* (Bristol: IOP Publishing)

[5] Blum K 2012 *Density Matrix Theory and Applications* (*Springer Series on Atomic, Optical and Plasma Physics*) 3rd edn (Heidelberg: Springer)

[6] Braunstein S L and van Loock P 2005 Quantum information with continuous variables *Rev. Mod. Phys.* **77** 513–77

[7] Davies R W and Davies K T R 1975 On the Wigner distribution function for an oscillator *Ann. Phys.* **89** 261–73

[8] Dirac P A M 1927 The quantum theory of the emission and absorption of radiation *Proc. R. Soc.* A **114** 243–65

[9] Dodd P J and Halliwell J J 2004 Disentanglement and decoherence by open system dynamics *Phys. Rev.* A **69**

[10] Doman B G S 2015 *The Classical Orthogonal Polynomials* (Singapore: World Scientific)

[11] Ekert A K and Knight P L 1989 Correlations and squeezing of two-mode oscillations *Am. J. Phys.* **57** 692–7

[12] Fano U 1957 Description of states in quantum mechanics by density matrix and operator techniques *Rev. Mod. Phys.* **29** 74–93

[13] Ferraro A, Olivares S, and Paris M G A 2005 *Gaussian States in Quantum Information* (*Napoli Series on Physics and Astrophysics*) (Brussels: Bibliopolis)

[14] Fetter A L and Walecka J D 2003 *Quantum Theory of Many-particle Systems* (Mineola, NY: Dover)

[15] Feynman R P 1998 *Statistical Mechanics: A Set of Lectures* (*Advanced Book Classics*) (Boulder, CO: Westview Press)

[16] Feynman R P, Kislinger M, and Ravndal F 1971 Current matrix elements from a relativistic quark model *Phys. Rev.* D **3** 2706–32

[17] Furusawa A and Van Loock P 2011 *Quantum Teleportation and Entanglement: A Hybrid Approach to Optical Quantum Information Processing* (Weinheim: Wiley)

[18] Ge W, Tasgin M E, and Zubairy M S 2015 Conservation relation of nonclassicality and entanglement for Gaussian states in a beam splitter *Phys. Rev.* A **92**

[19] Giedke G, Wolf M M, Krüger O, Werner R F, and Cirac J I 2003 Entanglement of formation for symmetric Gaussian states *Phys. Rev. Lett.* **91** 107901-1–4

[20] Gingrich R M and Adami C 2002 Quantum entanglement of moving bodies *Phys. Rev. Lett.* **89**

[21] Han D, Kim Y S, and Noz M E 1990 Lorentz-squeezed hadrons and hadronic temperature *Phys. Lett.* A **144** 111–5

[22] Han D, Kim Y S, and Noz M E 1999 Illustrative example of Feynman's rest of the Universe *Am. J. Phys.* **67** 61–6

[23] Kim M S, Son W, Bužek V, and Knight P L 2002 Entanglement by a beam splitter: nonclassicality as a prerequisite for entanglement *Phys. Rev.* A **65**

[24] Kim Y S and Li M 1989 Squeezed states and thermally excited states in the Wigner phase-space picture of quantum mechanics *Phys. Lett.* A **139** 445–8

[25] Kim Y S and Noz M E 1986 *Theory and Applications of the Poincaré Group* (Dordrecht: Springer)

[26] Kim Y S and Noz M E 1991 *Phase Space Picture of Quantum Mechanics: Group Theoretical Approach* (*Lecture Notes in Physics* vol 40) (Singapore: World Scientific)

[27] Kim Y S and Noz M E 2005 Coupled oscillators, entangled oscillators, and Lorentz–covariant harmonic oscillators *J. Opt.* B **7** S458–67

[28] Kim Y S and Noz M E 2012 Dirac matrices and Feynman rest of the Universe *Symmetry* **4** 626–43

[29] Kim Y S, Noz M E, and Oh S H 1979 A simple method for illustrating the difference between the homogeneous and inhomogeneous Lorentz groups *Am. J. Phys.* **47** 892–7

[30] Kim Y S and Wigner E P 1990 Entropy and Lorentz transformations *Phys. Lett.* A **147** 343–7

[31] Landau L D and Lifshitz E M 2008 *Statistical Physics, Part 1* (*Course of Theoretical Physics* vol 5) 3rd edn (Amsterdam: Elsevier)

[32] Leonhardt U 2010 *Essential Quantum Optics: From Quantum Measurements to Black Holes* (Cambridge: Cambridge University Press)

[33] Magnus W, Oberhettinger F, and Soni R P 1966 *Formulas and Theorems for the Special Functions of Mathematical Physics* (Berlin: Springer)

[34] Mann A and Revzen M 1989 Thermal coherent states *Phys. Lett.* A **134** 273–5

[35] Mufti A, Schmitt H A, and Sargent M 1993 Finite-dimensional matrix representations as calculational tools in quantum optics *Am. J. Phys.* **61** 729–33

[36] Paris M G A 1999 Entanglement and visibility at the output of a Mach–Zehnder interferometer *Phys. Rev.* A **59** 1615–21

[37] Rotbart F C 1981 Complete orthogonality relations for the covariant harmonic oscillator *Phys. Rev.* D **23** 3078–80

[38] Ruiz M J 1974 Orthogonality relation for covariant harmonic-oscillator wave functions *Phys. Rev.* D **10** 4306–7

[39] Umezawa H, Matsumoto H, and Tachiki M 1982 *Thermo Field Dynamics and Condensed States* (Amsterdam: North Holland)

[40] Von Neumann J, Beyer R T, and Wheeler N A 2018 *Mathematical Foundations of Quantum Mechanics* (Princeton, NJ: Princeton University Press)

[41] Walls D F and Milburn G J 2008 *Quantum Optics* 2nd edn (Berlin: Springer)

[42] Wigner E P 1932 On the quantum correction for thermodynamic equilibrium *Phys. Rev.* **40** 749–59

[43] Wigner E P and Yanase M M 1963 Information contents of distributions *Proc. Natl Acad. Sci.* **49** 910–8

[44] Yurke B and Potasek M 1987 Obtainment of thermal noise from a pure quantum state *Phys. Rev.* A **36** 3464–6

IOP Publishing

Mathematical Devices for Optical Sciences

Sibel Başkal, Young S Kim, and Marilyn E Noz

Chapter 9

Ray optics and optical activities

The Lorentz group, which serves as the basic language for Einstein's special theory of relativity, can also be considered to be the basic mathematical instrument in optical sciences, particularly in ray and polarization optics. As it is known that optical activities can perform rotations, we show that the rotation, if modulated by attenuations, can perform symmetry operations of Wigner's little groups. Conversely, the optical activity can serve as an analog computer for internal space–time symmetries of elementary particles.

As we saw in chapter 3, Wigner noted a particle can have internal variables in addition to the energy and momentum [22]. For instance, an electron can have spin degrees of freedom, in addition to momentum and energy. Photons can have helicity and gauge degrees of freedom. Wigner formulated this symmetry problem by introducing three-parameter subgroups of the Lorentz group, which preserve the four-momentum of a given particle.

For the discussion given here, let us assume that the optical ray propagates along the z-direction, and that the polarization rotates on the xy-plane. If the attenuation along the x-direction is different from that along the y-direction, this results in an asymmetric attenuation with a rotation around the z-axis. We will start with the group of two-by-two beam transfer matrices, commonly called the $ABCD$ matrices, which constitute a two-by-two representation of the Lorentz group. Thus it is possible to study this group in terms of the mathematical device developed for studying Wigner's little groups. In the real world, all optical rays go through attenuations. If the attenuation is axially symmetric, it does not raise additional mathematical problems. However, if the attenuation is anti-symmetric, there are further mathematical complications which arise. We show that the resulting mathematics not only allows analytical calculations of the optical activities with asymmetric attenuation effects, but also provides a computational instrument for Wigner's little group which dictates the internal space–time symmetries of elementary particles.

doi:10.1088/2053-2563/aafe78ch9

In section 9.1 we look first at ray optics in terms of the group of two-by-two *ABCD* matrices applicable to the Jones vector. It is not difficult to write matrices performing rotations and attenuations separately. We show how to equi-diagonalize the *ABCD* matrices and we look at some decompositions and recompositions of them. In section 9.2 we give some physical examples using the *ABCD* matrices applied to single lens and multilayer lens systems and camera optics. We show some relationships to particle physics. In section 9.3 we look at optical materials. It is a simple matter to construct a rotation matrix for a given value of propagation distance z and for an asymmetric attenuation. However, the problem becomes non-trivial when these two effects are combined at a microscopic scale with a small value of z. We compute the transformation matrix if the operation of rotation and squeeze are performed at a microscopic scale, and are accumulated to a macroscopic scale, thus providing a computational instrument for Wigner's little groups.

9.1 Ray optics using the group of *ABCD* matrices

Let us start with a light wave taking the form

$$\begin{pmatrix} E_x \\ E_y \end{pmatrix} = \begin{pmatrix} A \exp\left\{i(kz - \omega t + \phi_1)\right\} \\ B \exp\left\{i(kz - \omega t + \phi_2)\right\} \end{pmatrix}. \tag{9.1}$$

This ray can go through rotations around the z-axis. It can also go through xy asymmetric phase shifts and attenuations. The mathematics of these aspects is known as the Jones matrix formalism.

In particle physics, the Lorentz group is generated by six matrices. The elements of the *ABCD* matrices, however, are always real and unimodular (determinant = 1). As we saw in section 2.5, this group can be generated by J_2, K_3, and K_1, which lead to two-by-two matrices with real elements. These generators form a closed set of commutation relations given by

$$[J_2, K_1] = -iK_3, \quad [J_2, K_3] = iK_1, \quad \text{and} \quad [K_1, K_3] = -iJ_2, \tag{9.2}$$

where

$$J_2 = \frac{i}{2}\begin{pmatrix} 0 & -1 \\ 1 & 0 \end{pmatrix}, \quad K_3 = \frac{i}{2}\begin{pmatrix} 1 & 0 \\ 0 & -1 \end{pmatrix}, \quad \text{and} \quad K_1 = \frac{i}{2}\begin{pmatrix} 0 & 1 \\ 1 & 0 \end{pmatrix}. \tag{9.3}$$

These generators form the Lie algebra for the $Sp(2)$ group, as discussed in section 2.5 [6, 8, 21].

This representation of the $Sp(2)$ group consists of a rotation about the origin, a squeeze along the x-direction and another squeeze along axes rotated by 45°, respectively. Another representation of the $Sp(2)$ group which has two shear transformations and a squeeze transformation [2, 3] can also be obtained from these generators. As a result we will use in what follows a rotation, a squeeze, and a shear matrix of the form [4]

$$R(\theta) = \begin{pmatrix} \cos(\theta/2) & -\sin(\theta/2) \\ \sin(\theta/2) & \cos(\theta/2) \end{pmatrix}, \qquad B(\eta) = \begin{pmatrix} e^{\eta/2} & 0 \\ 0 & e^{-\eta/2} \end{pmatrix}, \qquad (9.4)$$

and

$$T(\gamma) = \begin{pmatrix} 1 & -\gamma \\ 0 & 1 \end{pmatrix}. \qquad (9.5)$$

The last matrix has also been equated to an optical filter [12], or a translation matrix [3].

It should be noted that the traces of these matrices are smaller than, greater than, and equal to 2, respectively. How to make transitions from one to another was discussed in chapter 3.

The *ABCD* matrices have three independent parameters because they have real elements and a determinant of 1. Optical materials and how they are arranged determine the elements of the *ABCD* matrices. We explore first the mathematical properties of the *ABCD* matrix which can address more fundamental issues in physics.

9.1.1 Diagonalization properties of the *ABCD* matrices

When using matrices, it is common practice to diagonalize them. This, however, is not always possible. For example, in equation (9.5), although the triangular matrix cannot be diagonalized, the two diagonal elements are equal.

In dealing with the *ABCD* matrices, we often encounter matrices which do not have equal diagonal elements, that is, they are not equi-diagonal. To make the diagonal elements equal it is possible to start from the matrix

$$[ABCD] = \begin{pmatrix} A & B \\ C & D \end{pmatrix}, \qquad (9.6)$$

where A and D are not necessarily equal to each other. Using the matrix $B(\eta)$ from equation (9.4) we apply the transformation

$$B(\eta)\ [ABCD]\ B(\eta) = \begin{pmatrix} e^{\eta/2} & 0 \\ 0 & e^{-\eta/2} \end{pmatrix} \begin{pmatrix} A & B \\ C & D \end{pmatrix} \begin{pmatrix} e^{\eta/2} & 0 \\ 0 & e^{-\eta/2} \end{pmatrix}. \qquad (9.7)$$

This yields the equi-diagonal form

$$\begin{pmatrix} \sqrt{AD} & B \\ C & \sqrt{AD} \end{pmatrix}, \qquad (9.8)$$

with

$$e^{\eta} = \sqrt{D/A}. \qquad (9.9)$$

We will use this form of equi-diagonalization when discussing camera optics in section 9.2.2.

It should be noted that the transformation of equation (9.7) is an unimodular (determinant-preserving) transformation, but it is not a similarity transformation. If we now consider a rotation of the $ABCD$ matrix

$$R(\theta)\ [ABCD]\ R(-\theta)$$
$$= \begin{pmatrix} \cos(\theta/2) & -\sin(\theta/2) \\ \sin(\theta/2) & \cos(\theta/2) \end{pmatrix} \begin{pmatrix} A & B \\ C & D \end{pmatrix} \begin{pmatrix} \cos(\theta/2) & \sin(\theta/2) \\ -\sin(\theta/2) & \cos(\theta/2) \end{pmatrix}, \qquad (9.10)$$

we obtain the matrix

$$\begin{pmatrix} A' & B' \\ C' & D' \end{pmatrix}, \qquad (9.11)$$

with

$$A' = \frac{1}{2}[A(1 + \cos\theta) + D(1 - \cos\theta) - (C + B)\sin\theta],$$
$$D' = \frac{1}{2}[A(1 - \cos\theta) + D(1 + \cos\theta) + (C + B)\sin\theta]. \qquad (9.12)$$

To make these two diagonal elements equal,

$$\tan\theta = \frac{A - D}{A + B}. \qquad (9.13)$$

This gives us two different ways of transforming $ABCD$ matrices to have equal diagonal elements. To make these different transformations into one mathematical transformation, we let M be an arbitrary element of the $Sp(2)$ group. Then we can define the *Hermitian transformation* of the $ABCD$ matrices as

$$M\ [ABCD]\ M^{\dagger}. \qquad (9.14)$$

Here $M^{\dagger}$ is the Hermitian conjugate of M.

We see that the Hermitian transformation of equation (9.14) is like the Lorentz transformation on the four-vector discussed in chapter 2. If it is to be a similarity transformation, then M must be Hermitian. This means that M must be anti-symmetric and, therefore, its Hermitian conjugate is its inverse. On the other hand, if M is symmetric, its Hermitian conjugate is not its inverse and the transformation is not a similarity transformation.

As the rotation matrix of equation (9.4) is anti-symmetric and hence Hermitian, the transformation of equation (9.14) with the rotation matrix is a similarity transformation. The squeeze matrix $B(\eta)$ is symmetric, hence invariant under Hermitian conjugation. This makes the the Hermitian transformation of equation (9.14) with the squeeze matrix not a similarity transformation.

When an $ABCD$ matrix has been equi-diagonalized, it can be brought to one of the forms [4, 5]

$$W'(\theta, \eta) = \begin{pmatrix} \cos(\theta/2) & -e^{\eta}\sin(\theta/2) \\ e^{-\eta}\sin(\theta/2) & \cos(\theta/2) \end{pmatrix},$$

$$W'(\gamma, \eta) = \begin{pmatrix} 1 & -e^{\eta}\gamma \\ 0 & 1 \end{pmatrix}, \tag{9.15}$$

$$W'(\lambda, \eta) = \begin{pmatrix} \cosh(\lambda/2) & e^{\eta}\sinh(\lambda/2) \\ e^{-\eta}\sinh(\lambda/2) & \cosh(\lambda/2) \end{pmatrix}.$$

We shall use the notation W' for these three equi-diagonal matrices, where the prime indicates that these are the Lorentz boosted version of Wigner matrices defined in table 3.4. It is then possible to use all the mathematical instruments developed for the two-by-two representation of Wigner's little groups in chapter 3. The expressions for these two-by-two Lorentz boosted matrices are given in table 9.1.

Like the matrices defined in equations (9.4) and (9.5), these three matrices form different classes with different traces which are smaller than, equal to, and greater than 2, respectively. As we saw in sections 3.4 and 3.5, it is possible to make continuations from one to the other through the tangential continuity.

9.1.2 Decompositions of the *ABCD* matrices

The equi-diagonal matrices of equation (9.15) can now be written as

$$B(\eta) \ W \ [B(\eta)]^{-1}, \tag{9.16}$$

where W is one of the three single-parameter matrices:

$$R(\theta) = \begin{pmatrix} \cos(\theta/2) & -\sin(\theta/2) \\ \sin(\theta/2) & \cos(\theta/2) \end{pmatrix},$$

$$T(\gamma) = \begin{pmatrix} 1 & -\gamma \\ 0 & 1 \end{pmatrix}, \tag{9.17}$$

$$S(\lambda) = \begin{pmatrix} \cosh(\lambda/2) & \sinh(\lambda/2) \\ \sinh(\lambda/2) & \cosh(\lambda/2) \end{pmatrix}.$$

Table 9.1. Lorentz boosted Wigner momenta and Wigner matrices in the two-by-two representation of the Lorentz group. This table is the two-by-two Lorentz boosted version of table 3.4.

Mass	Wigner momentum	Wigner matrix
Massive	$\begin{pmatrix} 1 & 0 \\ 0 & 1 \end{pmatrix}$	$\begin{pmatrix} \cos(\theta/2) & -e^{\eta}\sin(\theta/2) \\ e^{-\eta}\sin(\theta/2) & \cos(\theta/2) \end{pmatrix}$
Massless	$\begin{pmatrix} 1 & 0 \\ 0 & 0 \end{pmatrix}$	$\begin{pmatrix} 1 & -e^{\eta}\gamma \\ 0 & 1 \end{pmatrix}$
Imaginary mass	$\begin{pmatrix} 1 & 0 \\ 0 & -1 \end{pmatrix}$	$\begin{pmatrix} \cosh(\lambda/2) & e^{\eta}\sinh(\lambda/2) \\ e^{-\eta}\sinh(\lambda/2) & \cosh(\lambda 2) \end{pmatrix}$

These matrices, given in table 3.4, perform different physical operations. As we saw previously, the transformation of equation (9.16) by equation (9.4) is not a Hermitian transformation. It is a similarity transformation.

Since the two-by-two $ABCD$ matrices can be written as a similarity transformation of one of the three possible W matrices given in equation (9.17), they can be made equi-diagonalized by a rotation. The similarity transformation can be written as

$$[ABCD] = R(\sigma)B(\eta) \ W \ B(-\eta)R(-\sigma) = [R(\sigma)B(\eta)] \ W \ [R(\sigma)B(\eta)]^{-1}. \quad (9.18)$$

With repeated application of an $ABCD$ matrix this becomes

$$[ABCD]^N = [R(\sigma)B(\eta)] \ W^N \ [R(\sigma)B(\eta)]^{-1}. \quad (9.19)$$

Another form of decomposition known as the Bargmann decomposition was discussed in section 2.6.1. With this procedure we can combine all three different classes of equation (9.15) into one analytic expression as

$$W_B = R(\alpha)S(-2\chi)R(\alpha). \quad (9.20)$$

Here the forms of the rotation matrix R and the squeeze matrix S are given in equation (9.17). If we carry out the matrix multiplication, the W_B matrix becomes

$$\begin{pmatrix} (\cosh\chi)\cos\alpha & -\sinh\chi - (\cosh\chi)\sin\alpha \\ -\sinh\chi + (\cosh\chi)\sin\alpha & (\cosh\chi)\cos\alpha \end{pmatrix}, \quad (9.21)$$

where both α and χ are independent variables. These parameters can be written in terms of the η, θ, and γ for the matrices given in equation (9.15) by comparing the matrix elements.

If $(\cosh\chi)\sin\alpha > \sinh\chi$, the diagonal elements become

$$\cos(\theta/2) = (\cosh\chi)\cos\alpha. \quad (9.22)$$

The off-diagonal elements are then

$$e^{2\eta} = \frac{(\cosh\chi)\sin\alpha + \sinh\chi}{(\cosh\chi)\sin\alpha - \sinh\chi}. \quad (9.23)$$

Should the off-diagonal elements have the same sign, then the diagonal elements become

$$\cosh(\lambda/2) = (\cosh\chi)\cos\alpha, \quad (9.24)$$

which leads to

$$e^{2\eta} = \frac{(\cosh\chi)\sin\alpha + \sinh\chi}{\sinh\chi - (\cosh\chi)\sin\alpha}. \quad (9.25)$$

When the lower left element of equation (9.21) vanishes, we obtain the matrix

$$\begin{pmatrix} 1 & -2\sinh\chi \\ 0 & 1 \end{pmatrix}. \quad (9.26)$$

When we make a transition from one form to another among the three matrices given in equation (9.15), then the matrix of equation (9.21) is useful. This procedure is needed in discussing multilayer optics.

9.1.3 Recomposition of the *ABCD* matrices

As to the recomposition of the *ABCD* matrices, the best physical example for its demonstration can be derived from para-axial lens optics, where the lens and translation matrices take the form

$$L(f) = \begin{pmatrix} 1 & 0 \\ -1/f & 1 \end{pmatrix} \quad \text{and} \quad T(d) = \begin{pmatrix} 1 & d \\ 0 & 1 \end{pmatrix}. \tag{9.27}$$

Here f is the focal length of the lens and d is the distance between lenses. They are applicable to the two-dimensional space of $(y, m)^T$, where y measures the height of the ray from the propagation axis and m is the slope of the ray. Both L and T convert multiplications into additions

$$L(f_1 + f_2) = L(f_1) L(f_2) \quad \text{and} \quad T(d_1 + d_2) = T(d_1) T(d_2) \tag{9.28}$$

if computations are restricted to L-type or T-type matrices.

A one-lens system consists of a *TLT* chain, while a two-lens system has the form *TLTLT*. The chain becomes longer when more lenses are included. However, the net result will be one *ABCD* matrix with three independent real parameters. As mentioned earlier, this is a representation of the $Sp(2)$ group. Its most general form can be expressed as $W''(\phi, \rho, \eta) = R(\phi)B(\eta)R(\rho)$ or more explicitly as

$$W''(\phi, \rho, \eta) = \begin{pmatrix} \cos\phi & -\sin\phi \\ \sin\phi & \cos\phi \end{pmatrix} \begin{pmatrix} e^{\eta} & 0 \\ 0 & e^{-\eta} \end{pmatrix} \begin{pmatrix} \cos\rho & -\sin\rho \\ \sin\rho & \cos\rho \end{pmatrix}, \tag{9.29}$$

by considering the Bargmann decomposition given in section 2.6.1. Let us note that $R(\rho)$ can be decomposed as $R(\rho) = R(-\phi)R(\theta)$, where $R(\theta)$ is as in table 2.1. Thus in terms of matrices, the last matrix in equation (9.29) becomes

$$R(\rho) = \begin{pmatrix} \cos\phi & \sin\phi \\ -\sin\phi & \cos\phi \end{pmatrix} \begin{pmatrix} \cos\theta & -\sin\theta \\ \sin\theta & \cos\theta \end{pmatrix}, \tag{9.30}$$

with $\rho = \theta - \phi$. Finally, we have

$$W''(\phi, \rho, \eta) = R(\phi)B(\eta)R(-\phi)R(\theta) \tag{9.31}$$

and instead of the angle ρ, now θ is an independent parameter.

The matrix W'' can be obtained from the product of two matrices, one symmetric S and the other orthogonal R as $W''(\phi, \theta, \eta) = S(\phi, \eta)R(\theta)$, with $S(\phi, \eta) = R(\phi)B(\eta)R(-\phi)$. This symmetric matrix S takes the form [12]

$$S(\phi, \eta) = \begin{pmatrix} \cosh\eta + (\sinh\eta)\cos(2\phi) & (\sinh\eta)\sin(2\phi) \\ (\sinh\eta)\sin(2\phi) & \cosh\eta - (\sinh\eta)\cos(2\phi) \end{pmatrix}. \tag{9.32}$$

Our procedure is to write S and R separately as L and T chains. Let us consider first the rotation matrix that can be obtained through the multiplication of L-type and T-type matrices as [17]

$$R(\theta) = \begin{pmatrix} 1 & -\tan(\theta/2) \\ 0 & 1 \end{pmatrix} \begin{pmatrix} 1 & 0 \\ \sin\theta & 1 \end{pmatrix} \begin{pmatrix} 1 & -\tan(\theta/2) \\ 0 & 1 \end{pmatrix}. \tag{9.33}$$

This expression is in the form of TLT, but it can also be written in the form of LTL. If we take the transpose and change the sign of θ, R becomes

$$R'(\theta) = \begin{pmatrix} 1 & 0 \\ \tan(\theta/2) & 1 \end{pmatrix} \begin{pmatrix} 1 & -\sin\theta \\ 0 & 1 \end{pmatrix} \begin{pmatrix} 1 & 0 \\ \tan(\theta/2) & 1 \end{pmatrix}. \tag{9.34}$$

Both R and R' are the same matrix but are decomposed in different ways.

As for the two-parameter symmetric matrix of equation (9.32), we start with the form $LTLT$ [2],

$$S(a, b) = \begin{pmatrix} 1 & 0 \\ b & 1 \end{pmatrix} \begin{pmatrix} 1 & a \\ 0 & 1 \end{pmatrix} \begin{pmatrix} 1 & 0 \\ a & 1 \end{pmatrix} \begin{pmatrix} 1 & b \\ 0 & 1 \end{pmatrix}, \tag{9.35}$$

which can be combined into one symmetric matrix:

$$S(a, b) = \begin{pmatrix} 1 + a^2 & b(1 + a^2) + a \\ b(1 + a^2) + a & 1 + 2ab + b^2(1 + a^2) \end{pmatrix}. \tag{9.36}$$

By comparing equations (9.32) and (9.36), we can compute the parameters a and b in terms of η and ϕ. The result is

$$a = \pm\sqrt{(\cosh\eta - 1) + (\sinh\eta)\cos(2\phi)}\,,$$
$$b = \frac{(\sinh\eta)\sin(2\phi) \mp \sqrt{(\cosh\eta - 1) + (\sinh\eta)\cos(2\phi)}}{\cosh\eta + (\sinh\eta)\cos(2\phi)}. \tag{9.37}$$

This matrix can also be written in a $TLTL$ form:

$$S'(a', b') = \begin{pmatrix} 1 & b' \\ 0 & 1 \end{pmatrix} \begin{pmatrix} 1 & 0 \\ a' & 1 \end{pmatrix} \begin{pmatrix} 1 & a' \\ 0 & 1 \end{pmatrix} \begin{pmatrix} 1 & 0 \\ b' & 1 \end{pmatrix}. \tag{9.38}$$

Then the parameters a' and b' are

$$a' = \pm\sqrt{(\cosh\eta - 1) - (\sinh\eta)\cos(2\phi)}\,,$$
$$b' = \frac{(\sinh\eta)\sin(2\phi) \mp \sqrt{(\cosh\eta - 1) - (\sinh\eta)\cos(2\phi)}}{\cosh\eta - (\sinh\eta)\cos(2\phi)}. \tag{9.39}$$

The difference between the two sets of parameters ab and $a'b'$ is the sign of the parameter η. This sign change means that the squeeze operation is in the direction perpendicular to the original direction. In choosing ab or $a'b'$, we will also have to take care of the sign of the quantity inside the square root to be positive. If $\cos(2\phi)$ is sufficiently small, both sets are acceptable. On the other hand, if the absolute value of $(\sinh \eta) \cos(2\phi)$ is greater than $(\cosh \eta - 1)$, only one of the sets, ab or $a'b'$, is valid.

We can now combine S and R matrices in order to construct the $ABCD$ matrix. In so doing, the number of matrices is reduced by one as

$$SR = \begin{pmatrix} 1 & 0 \\ b & 1 \end{pmatrix}\begin{pmatrix} 1 & a \\ 0 & 1 \end{pmatrix}\begin{pmatrix} 1 & 0 \\ a & 1 \end{pmatrix}\begin{pmatrix} 1 & b - \tan(\theta/2) \\ 0 & 1 \end{pmatrix}$$
$$\times \begin{pmatrix} 1 & 0 \\ \sin\theta & 1 \end{pmatrix}\begin{pmatrix} 1 & -\tan(\theta/2) \\ 0 & 1 \end{pmatrix}. \tag{9.40}$$

We can also combine making the product $S'R'$, which becomes

$$S'R' = \begin{pmatrix} 1 & b' \\ 0 & 1 \end{pmatrix}\begin{pmatrix} 1 & 0 \\ a' & 1 \end{pmatrix}\begin{pmatrix} 1 & a' \\ 0 & 1 \end{pmatrix}\begin{pmatrix} 1 & 0 \\ b' + \tan(\theta/2) & 1 \end{pmatrix}$$
$$\times \begin{pmatrix} 1 & -\sin\theta \\ 0 & 1 \end{pmatrix}\begin{pmatrix} 1 & 0 \\ \tan(\theta/2) & 1 \end{pmatrix}. \tag{9.41}$$

To reduce SR of equation (9.40), two adjoining T matrices were combined into one T matrix. Similarly, two L matrices were combined into one, to simplify $S'R'$ of equation (9.41).

In both cases, there are six matrices, consisting of three T and three L matrices. Now, we have come to the conclusion that the minimum number of L- and T-type matrices required for the the most general form of the $ABCD$ matrix is six.

In para-axial optics, we often encounter special forms of the $ABCD$ matrix. For instance, the squeeze matrix $B(2\eta)$ in equation (9.29) is for pure magnification [10]. This is a special case of the decomposition given for S and S' in equations (9.36) and (9.38), respectively, with $\phi = 0$. However, if η is positive, the set $a'b'$ is not acceptable because the quantity in the square root in equation (9.39) becomes negative. For the ab set

$$a = \pm(e^\eta - 1)^{1/2} \quad \text{and} \quad b = \mp e^{-\eta}(e^\eta - 1)^{1/2}. \tag{9.42}$$

The L and T matrices of equation (9.27) are products of Iwasawa decompositions as given in section 2.6.2. Since the most general form of the $ABCD$ matrix is obtained by implementing the inverse process of the Iwasawa decomposition, using solely the products of L and T matrices, we have the recomposition of the $ABCD$ matrix.

9.2 Physical examples using $ABCD$ matrices

There are many periodic systems. To illustrate the use of the $ABCD$ matrices we will discuss three examples. First we will consider a laser cavity consisting of two identical concave mirrors separated by a distant d as shown in figure 9.1.

(a)

The cycle can start anywhere
between the lenses.

Convex lens

z

The cycle starts at the midway between
the lenses when [ABCD] is equi-diagonal.

(b)

Concave mirror surface

midway

$\leftarrow$(x)$\rightarrow$

$\leftarrow$ (d/2) $\rightarrow$

z

$\leftarrow$ (d - x) $\rightarrow$

d

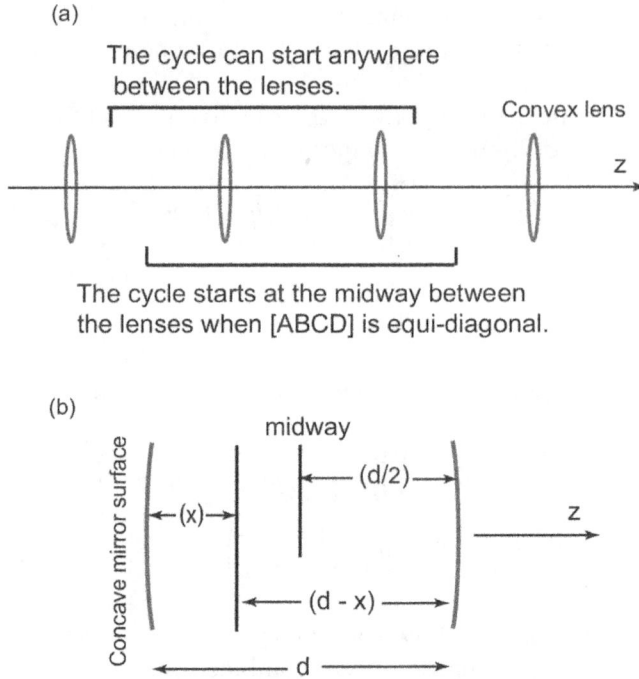

Figure 9.1. Optical rays in the laser cavity: (a) illustrates that reflected beams are like those going though a series of convex lenses and (b) describes how the beams are reflected from concave mirror surfaces [6].

A round trip of one beam can be expressed by the $ABCD$ matrix as

$$[ABCD] = \begin{pmatrix} 1 & 0 \\ -2/R & 1 \end{pmatrix} \begin{pmatrix} 1 & d \\ 0 & 1 \end{pmatrix} \begin{pmatrix} 1 & 0 \\ -2/R & 1 \end{pmatrix} \begin{pmatrix} 1 & d \\ 0 & 1 \end{pmatrix}. \tag{9.43}$$

Here the matrices

$$L(R) = \begin{pmatrix} 1 & 0 \\ -2/R & 1 \end{pmatrix} \quad \text{and} \quad T(d) = \begin{pmatrix} 1 & d \\ 0 & 1 \end{pmatrix} \tag{9.44}$$

are the mirror and translation matrices, respectively, where R is the radius of the mirror and d is the mirror separation. This form is quite familiar to us from the laser literature [13, 14, 23].

What then happens when this process is repeated? This leads to the question of whether the chain of matrices in equation (9.43) can be brought to an equi-diagonal form. Let us rewrite the matrix of equation (9.43) as

$$[ABCD] = \begin{pmatrix} 1 & -d/2 \\ 0 & 1 \end{pmatrix} \begin{pmatrix} 1 & d/2 \\ 0 & 1 \end{pmatrix} \begin{pmatrix} 1 & 0 \\ -2/R & 1 \end{pmatrix} \begin{pmatrix} 1 & d/2 \\ 0 & 1 \end{pmatrix}^2$$
$$\times \begin{pmatrix} 1 & 0 \\ -2/R & 1 \end{pmatrix} \begin{pmatrix} 1 & d/2 \\ 0 & 1 \end{pmatrix} \begin{pmatrix} 1 & d/2 \\ 0 & 1 \end{pmatrix}. \tag{9.45}$$

This translates the system by $-d/2$ using a translation matrix given in equation (9.44). The *ABCD* matrix of equation (9.43) can then be written as

$$[ABCD] = \begin{pmatrix} 1 & -d/2 \\ 0 & 1 \end{pmatrix} \left[\begin{pmatrix} 1 - d/R & d - d^2/2R \\ -2/R & 1 - d/R \end{pmatrix} \right]^2 \begin{pmatrix} 1 & d/2 \\ 0 & 1 \end{pmatrix}. \tag{9.46}$$

Let us now concentrate on the matrix in the middle to obtain

$$\begin{pmatrix} 1 - d/R & d - d^2/2R \\ -2/R & 1 - d/R \end{pmatrix}, \tag{9.47}$$

which can be written as

$$\begin{pmatrix} \sqrt{d} & 0 \\ 0 & 1/\sqrt{d} \end{pmatrix} \begin{pmatrix} 1 - d/R & 1 - d/2R \\ -2d/R & 1 - d/R \end{pmatrix} \begin{pmatrix} 1/\sqrt{d} & 0 \\ 0 & \sqrt{d} \end{pmatrix}. \tag{9.48}$$

Now, the *ABCD* matrix can be decomposed into

$$[ABCD] = E \; C^2 \; E^{-1}, \tag{9.49}$$

where

$$C = \begin{pmatrix} 1 - d/R & 1 - d/2R \\ -2d/R & 1 - d/R \end{pmatrix} \tag{9.50}$$

and

$$E = \begin{pmatrix} 1 & -d/2 \\ 0 & 1 \end{pmatrix} \begin{pmatrix} \sqrt{d} & 0 \\ 0 & 1/\sqrt{d} \end{pmatrix}. \tag{9.51}$$

Since the *C* matrix contains only dimensionless numbers, we rewrite it as

$$C = \begin{pmatrix} \cos(\gamma/2) & e^\eta \sin(\gamma/2) \\ -e^{-\eta} \sin(\gamma/2) & \cos(\gamma/2) \end{pmatrix}, \tag{9.52}$$

with

$$\cos(\gamma/2) = 1 - \frac{d}{R},$$

$$e^\eta = \sqrt{\frac{2R - d}{4d}}. \tag{9.53}$$

In this case d and R are both positive, and the restriction on them is that d be smaller than $2R$. This stability condition is frequently mentioned in the literature [13, 14].

The equi-diagonal matrix C^2 becomes

$$\begin{pmatrix} \cos(\gamma) & e^\eta \sin(\gamma) \\ -e^{-\eta} \sin(\gamma) & \cos(\gamma) \end{pmatrix}, \tag{9.54}$$

and now attains the familiar form

$$B(\eta)R(2\gamma)B(-\eta). \tag{9.55}$$

This form is the same as that of the Wigner decomposition for particle physics where a particle is Lorentz boosted to the rest frame, rotated without changing the momentum, and then Lorentz boosted back to the original state. Hence, the $ABCD$ matrix has the form

$$[ABCD] = [E\ B(\eta)]\ R(2\gamma)\ [E\ B(\eta)]^{-1}. \tag{9.56}$$

Repeated application of this process allows us to write

$$[ABCD]^N = [E\ B(\eta)]\ R(2N\gamma)\ [E\ B(\eta)]^{-1}. \tag{9.57}$$

9.2.1 Optics using multilayers

When an optical beam goes through a periodic medium with two different refractive indexes, we know that when the beam traveling in the first medium hits the second medium, it is partially transmitted and partially reflected. Since it is necessary to maintain the continuity of the Poynting vector we define the electric fields for the optical beams in the first and second media, respectively, as

$$E_1^{(\pm)} = \frac{1}{\sqrt{n_1}} \exp\left(\pm ik_1 z - \omega t\right),$$
$$E_2^{(\pm)} = \frac{1}{\sqrt{n_2}} \exp\left(\pm ik_2 z - \omega t\right). \tag{9.58}$$

The incoming and reflected rays are represented by the superscripts $(+)$ and $(-)$, respectively.

The two-by-two $ABCD$ matrix relates the two optical rays by

$$\begin{pmatrix} E_2^{(+)} \\ E_2^{(-)} \end{pmatrix} = \begin{pmatrix} A & B \\ C & D \end{pmatrix} \begin{pmatrix} E_1^{(+)} \\ E_1^{(-)} \end{pmatrix}. \tag{9.59}$$

The transmission coefficients and the phase shifts the beams experience while going through the media determine the elements of the $ABCD$ matrix [1, 8].

As the beam passes through the first medium to the second, it is possible to use the boundary matrix given by Azzam and Bashara [1] and by [7, 18, 19]. We can write the matrix

$$S(\sigma) = \begin{pmatrix} \cosh(\sigma/2) & \sinh(\sigma/2) \\ \sinh(\sigma/2) & \cosh(\sigma/2) \end{pmatrix}, \tag{9.60}$$

where the σ parameter is given by

$$\cosh\left(\frac{\sigma}{2}\right) = \frac{n_1 + n_2}{2\sqrt{n_1 n_2}} \quad \text{and} \quad \sinh\left(\frac{\sigma}{2}\right) = \frac{n_1 - n_2}{2\sqrt{n_1 n_2}}. \tag{9.61}$$

Here n_1 and n_2 are the refractive indexes of the first and second medium, respectively. From figure 9.2 it can be seen that the boundary matrix for the beam going from the second medium should be $S(-\sigma)$.

Additionally, the phase shifts which the beams undergo as they travel through the medium have to be considered. For the beam going through the first medium, the phase-shift matrix can be written as

$$P(\delta_1) = \begin{pmatrix} e^{-i\delta_1/2} & 0 \\ 0 & e^{i\delta_1/2} \end{pmatrix}. \tag{9.62}$$

We get a similar expression for $P(\delta_2)$ for the second medium. The wave number and thickness of the medium determine the phase shift δ.

If we choose to consider one complete cycle starting from the midpoint of the second medium, we can write

$$M_c = P(\delta_2/2)S(\sigma)P(\delta_1)S(-\sigma)P(\delta_2/2). \tag{9.63}$$

When this is multiplied into one matrix, is this matrix equi-diagonal so that we can use the Wigner and Bargmann decompositions? We might also ask whether the matrices in the above expression can be converted into matrices with real elements.

To answer the second question, we consider the similarity transformation

$$C_1 \; P(\delta)S(\sigma) \; C_1^{-1}, \tag{9.64}$$

with

$$C_1 = \frac{1}{\sqrt{2}}\begin{pmatrix} 1 & i \\ i & 1 \end{pmatrix}. \tag{9.65}$$

Thus the result of this transformation can then be written

$$R(\delta)S(\sigma), \tag{9.66}$$

where

$$R(\delta) = \begin{pmatrix} \cos(\delta/2) & -\sin(\delta/2) \\ \sin(\delta/2) & \cos(\delta/2) \end{pmatrix}. \tag{9.67}$$

It can be seen that this notation is consistent with the rotation matrices used in section 9.1.2.

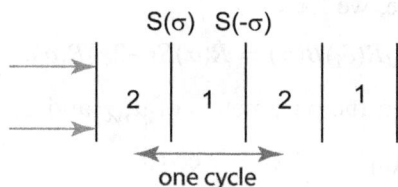

Figure 9.2. Multilayers. It is possible to start the cycle in such a way that the ABCD matrix becomes equi-diagonal [8, 9].

We can also make another similarity transformation with

$$C_2 = \frac{1}{\sqrt{2}}\begin{pmatrix} 1 & 1 \\ -1 & 1 \end{pmatrix}. \tag{9.68}$$

Here $S(\sigma)$ is changed into $B(\sigma)$ without changing $R(\delta)$, where

$$B(\sigma) = \begin{pmatrix} e^{\sigma/2} & 0 \\ 0 & e^{-\sigma/2} \end{pmatrix}, \tag{9.69}$$

again consistent with the $B(\eta)$ matrix used in section 9.1.2.

As a result the net similarity transformation matrix is [8]

$$C = C_2 C_1 = \frac{1}{\sqrt{2}}\begin{pmatrix} e^{i\pi/4} & e^{i\pi/4} \\ -e^{-i\pi/4} & e^{-i\pi/4} \end{pmatrix}, \tag{9.70}$$

with

$$C^{-1} = \frac{1}{\sqrt{2}}\begin{pmatrix} e^{-i\pi/4} & -e^{i\pi/4} \\ e^{-i\pi/4} & e^{i\pi/4} \end{pmatrix}. \tag{9.71}$$

If we apply this similarity transformation to the long matrix chain of equation (9.63), it becomes another chain

$$M_c' = R(\delta_2/2)B(\sigma)R(\delta_1)B(-\sigma)R(\delta_2/2), \tag{9.72}$$

where all the matrices are real.

The main question is whether this matrix chain can be brought to one equi-diagonal matrix. We can see that the three middle matrices can be written in a familiar form

$$M = B(\sigma)R(\delta_1)B(-\sigma)$$
$$= \begin{pmatrix} \cos(\delta_1/2) & -e^{\sigma}\sin(\delta_1/2) \\ e^{-\sigma}\sin(\delta_1/2) & \cos(\delta_1/2) \end{pmatrix}. \tag{9.73}$$

Because the rotation matrix $R(\delta_2/2)$ is at the beginning and at the end of equation (9.72), it is not clear whether the entire chain can be written as a similarity transformation.

To find a solution to this issue, we write M in equation (9.73) in the form of Bargmann decomposition W_B as in equation (9.20) whose explicit form is given in equation (9.21). Therefore, we have

$$B(\alpha)R(\delta_1)B(\alpha) = R(\alpha)S(-2\chi)R(\alpha), \tag{9.74}$$

where the relation between the parameters of α, χ and α, δ_1 is found to be

$$\cos(\delta_1/2) = (\cosh\chi)\cos\alpha,$$
$$e^{2\sigma} = \frac{(\cosh\chi)\sin\alpha + \sinh\chi}{(\cosh\chi)\sin\alpha - \sinh\chi}. \tag{9.75}$$

The entire chain M_c of equation (9.63) can clearly be written as another Bargmann decomposition

$$M_c'' = R(\alpha + \delta_2/2)S(-2\chi)R(\alpha + \delta_2/2). \tag{9.76}$$

It is possible now to convert this expression to another Wigner decomposition [9] by considering

$$B(\eta)R(\theta)B(-\eta). \tag{9.77}$$

Now we equate

$$R(\alpha + \delta_2/2)S(-2\chi)R(\alpha + \delta_2/2) = B(\eta)R(\theta)B(-\eta), \tag{9.78}$$

to obtain the relation between the parameters as

$$\cos(\theta/2) = (\cosh\chi)\cos(\alpha + \delta_2/2),$$
$$e^{2\eta} = \frac{(\cosh\chi)\sin(\alpha + \delta_2/2) + \sinh\chi}{(\cosh\chi)\sin(\alpha + \delta_2/2) - \sinh\chi}. \tag{9.79}$$

Now, we see that the decomposition of equation (9.77) allows us to deal with the periodic system of multilayers. For repeated application of M_c, we can now write

$$M_c^N = B(\eta)R(N\theta)B(-\eta). \tag{9.80}$$

9.2.2 Ray optics applied to cameras

A lens with focal length f and the propagation of the ray by an amount d [20] comprise the basic optical arrangement for a camera. We can write the lens and the ray translation matrices as

$$L(f) = \begin{pmatrix} 1 & 0 \\ -1/f & 1 \end{pmatrix} \quad \text{and} \quad T(d) = \begin{pmatrix} 1 & d \\ 0 & 1 \end{pmatrix}. \tag{9.81}$$

Suppose the object and the image are d_1 and d_2 distances away from the lens, respectively, then we can describe the system by an $ABCD$ matrix of the form

$$[ABCD] = \begin{pmatrix} 1 & d_2 \\ 0 & 1 \end{pmatrix}\begin{pmatrix} 1 & 0 \\ -1/f & 1 \end{pmatrix}\begin{pmatrix} 1 & d_1 \\ 0 & 1 \end{pmatrix}. \tag{9.82}$$

If we multiply the three matrices of equation (9.82) together the camera matrix takes the form

$$[ABCD] = \begin{pmatrix} 1 - d_2/f & d_1 + d_2 - d_1d_2/f \\ -1/f & 1 - d_1/f \end{pmatrix}. \tag{9.83}$$

A focused image is obtained when the upper-right element of this matrix vanishes [3] leading to

$$\frac{1}{d_1} + \frac{1}{d_2} = \frac{1}{f}. \tag{9.84}$$

Since both d_1 and d_2 must be longer than f for camera optics, it is more convenient to deal with the negative of the matrix given in equation (9.83) with positive diagonal elements. Additionally, we will use the dimensionless variables

$$x_1 = d_1/f \qquad \text{and} \qquad x_2 = d_2/f, \tag{9.85}$$

so that the camera matrix becomes

$$\begin{pmatrix} x_2 - 1 & (x_1 - 1)(x_2 - 1) - 1 \\ 1 & x_1 - 1 \end{pmatrix}. \tag{9.86}$$

Here we have applied the focal condition

$$\frac{1}{x_1} + \frac{1}{x_2} = 1. \tag{9.87}$$

To study the formula of the camera matrix equation (9.86) as a representation of the Lorentz group, the matrix must first be brought equi-diagonal form. In order to preserve the focal condition of equation (9.87), the off-diagonal elements should remain invariant. To do this it is necessary to resort to the Hermitian transformation given in equation (9.7) using the transformation matrix

$$\begin{pmatrix} e^{\varsigma/2} & 0 \\ 0 & e^{-\varsigma/2} \end{pmatrix}, \tag{9.88}$$

where

$$e^{-\varsigma} = \sqrt{\frac{1 - x_2}{1 - x_1}}. \tag{9.89}$$

Now, the diagonal elements are found to be

$$\sqrt{(1 - x_1)(1 - x_2)}. \tag{9.90}$$

For diagonal elements smaller than one, the camera matrix should be given by

$$\begin{pmatrix} \cos(\theta/2) & -e^{\eta}\sin(\theta/2) \\ e^{-\eta}\sin(\theta/2) & \cos(\theta/2) \end{pmatrix}, \tag{9.91}$$

with

$$e^{-\eta}\sin(\theta/2) = 1. \tag{9.92}$$

This results in the matrix

$$\begin{pmatrix} \cos(\theta/2) & -\sin^2(\theta/2) \\ 1 & \cos(\theta/2) \end{pmatrix}. \tag{9.93}$$

For the diagonal elements greater than one, the camera matrix should be given by

$$\begin{pmatrix} \cosh(\lambda/2) & e^{\eta}\sinh(\lambda/2) \\ e^{-\eta}\sinh(\lambda/2) & \cosh(\lambda/2) \end{pmatrix}, \tag{9.94}$$

with

$$e^{-\eta}\sinh(\lambda/2) = 1. \tag{9.95}$$

This leads to

$$\begin{pmatrix} \cosh(\lambda/2) & \sinh^2(\lambda/2) \\ 1 & \cosh(\lambda/2) \end{pmatrix}. \tag{9.96}$$

The focusing process produces a transition from

$$\begin{pmatrix} 1 - \xi^2/2 & -\xi^2 \\ 1 & 1 - \xi^2/2 \end{pmatrix} \text{ to } \begin{pmatrix} 1 + \xi^2/2 & \xi^2 \\ 1 & 1 + \xi^2/2 \end{pmatrix}, \tag{9.97}$$

via $\xi = 0$. Indeed, this is the tangential continuity discussed extensively in chapter 3.

9.3 Optical activities

Optical activities can be described by the real rotation and squeeze matrices given in equation (9.4). We assume here that the optical ray propagates along the z-direction, and that the polarization rotates on the xy-plane. If the attenuation along the x-direction is different from that along the y-direction, this results in an asymmetric attenuation and the rotation around the z-axis.

Thus as the ray propagates along the z-direction, the polarization goes through the rotation

$$R(\chi z) = \begin{pmatrix} \cos(\chi z) & -\sin(\chi z) \\ \sin(\chi z) & \cos(\chi z) \end{pmatrix}. \tag{9.98}$$

Here we have explicitly included the z-direction and for simplicity we use χ for $\theta/2$. This matrix is applicable to the x- and y-components of the polarization, and the rotation angle increases as z increases.

The optical ray is usually attenuated due to absorption by the medium through which it travels. The attenuation coefficient, however, could be different in each of the transverse directions. Hence if the rate of attenuation along the x-direction is different from that along the y-axis, this asymmetric attenuation can be described by the squeeze matrix, where we use μ for $\eta/2$:

$$\begin{pmatrix} e^{-\mu_1 z} & 0 \\ 0 & e^{-\mu_2 z} \end{pmatrix} = e^{-\lambda z}\begin{pmatrix} e^{\mu z} & 0 \\ 0 & e^{-\mu z} \end{pmatrix}, \tag{9.99}$$

with

$$\lambda = \frac{\mu_2 + \mu_1}{2},$$
$$\mu = \frac{\mu_2 - \mu_1}{2}. \tag{9.100}$$

The overall attenuation is given by the exponential factor $e^{-\lambda z}$. The matrix

$$\begin{pmatrix} e^{\mu z} & 0 \\ 0 & e^{-\mu z} \end{pmatrix}, \tag{9.101}$$

performs a squeeze transformation where the x-component of the polarization is expanded, while the y-component is contracted. This is therefore a squeeze along the x-direction.

The squeeze, however, can be in the direction which makes an angle ϕ with the x-axis and not along the x-direction. Let us call this matrix S. The squeeze matrix then becomes

$$S(\phi, \mu z) = \begin{pmatrix} \cos\phi & -\sin\phi \\ \sin\phi & \cos\phi \end{pmatrix}\begin{pmatrix} e^{\mu z} & 0 \\ 0 & e^{-\mu z} \end{pmatrix}\begin{pmatrix} \cos\phi & \sin\phi \\ -\sin\phi & \cos\phi \end{pmatrix}, \tag{9.102}$$

which can be compressed to one matrix

$$\begin{pmatrix} \cosh(\mu z) + \cos(2\phi)\sinh(\mu z) & \sin(2\phi)\sinh(\mu z) \\ \sin(2\phi)\sinh(\mu z) & \cosh(\mu z) - \cos(2\phi)\sinh(\mu z) \end{pmatrix}. \tag{9.103}$$

If $\phi = 45°$, this matrix becomes

$$S(\pi/4, \mu z) = \begin{pmatrix} \cosh(\mu z) & \sinh(\mu z) \\ \sinh(\mu z) & \cosh(\mu z) \end{pmatrix}. \tag{9.104}$$

In the following discussion, this form of squeeze matrix will be used and written as $S(\mu z)$ without the angle as is done in the above expression. Thus, if the squeeze is made along the x-axis, the squeeze matrix is

$$S(0, \mu z) = R(-\pi/4, \mu z)S(\mu z)R(\pi/4, \mu z). \tag{9.105}$$

If this squeeze is followed the rotation of equation (9.98), the net effect is

$$e^{-\lambda z}\begin{pmatrix} \cos(\chi z) & -\sin(\chi z) \\ \sin(\chi z) & \cos(\chi z) \end{pmatrix}\begin{pmatrix} \cosh(\mu z) & \sinh(\mu z) \\ \sinh(\mu z) & \cosh(\mu z) \end{pmatrix}, \tag{9.106}$$

where z is considered to be on a macroscopic scale, maybe measured in centimeters. However, this is not an accurate description of the optical process.

This optical process happens on a microscopic scale of z/N where N is a very large number. This becomes accumulated into the macroscopic scale of z after N repetitions. As we saw in section 9.2.1 the transformation matrix takes the same form as that given in equation (9.80), where

$$M(\chi, \mu, z) = [e^{-\lambda z/N} S(\mu z/N) R(\chi z/N)]^N. \tag{9.107}$$

In the limit of large N, this quantity becomes

$$e^{-\lambda z}\left[\begin{pmatrix} 1 & \mu z/N \\ \mu z/N & 1 \end{pmatrix}\begin{pmatrix} 1 & -\chi z/N \\ \chi z/N & 1 \end{pmatrix}\right]^N. \tag{9.108}$$

Since $\chi z/N$ and $\mu z/N$ are very small,

$$M(\chi, \mu, z) = e^{-\lambda z}\left[\begin{pmatrix} 1 & 0 \\ 0 & 1 \end{pmatrix} + \begin{pmatrix} 0 & -(\chi - \mu) \\ (\chi + \mu) & 0 \end{pmatrix}\frac{z}{N}\right]^N. \tag{9.109}$$

For large N, this matrix can be written as

$$M(\chi, \mu, z) = e^{-\lambda z}e^{Uz}, \tag{9.110}$$

with

$$U = \begin{pmatrix} 0 & -(\chi - \mu) \\ (\chi + \mu) & 0 \end{pmatrix}. \tag{9.111}$$

We now are interested in calculating the exponential form e^{Uz} by making a Taylor expansion. To do this we need to compute U^N. This is not a difficult problem if U is diagonal or can be diagonalized by a similarity transformation of a diagonal matrix. If this form cannot be diagonalized then a non-trivial problem arises.

9.3.1 Computation of the transformation matrix U

It is now necessary to compute the exponential form of equation (9.111). We consider three different possibilities.

If χ in equation (9.111) is greater than μ, the off-diagonal elements have opposite signs, and we can write U as

$$U = k\begin{pmatrix} 0 & -e^{2\zeta} \\ e^{-2\zeta} & 0 \end{pmatrix}, \tag{9.112}$$

with

$$k = \sqrt{\chi^2 - \mu^2},$$
$$e^{2\zeta} = \sqrt{\frac{\chi + \mu}{\chi - \mu}}, \tag{9.113}$$

or conversely

$$\chi = k \cosh(2\zeta) \quad \text{and} \quad \mu = k \sinh(2\zeta). \tag{9.114}$$

If μ is greater than χ, the off-diagonal elements have the same sign. We can then write U as

$$U = k\begin{pmatrix} 0 & e^{2\zeta} \\ e^{-2\zeta} & 0 \end{pmatrix}, \tag{9.115}$$

with

$$k = \sqrt{\mu^2 - \chi^2},$$
$$e^{2\zeta} = \sqrt{\frac{\mu + \chi}{\mu - \chi}}, \tag{9.116}$$

or

$$\chi = k \sinh(2\zeta) \quad \text{and} \quad \mu = k \cosh(2\zeta). \tag{9.117}$$

If $\chi = \mu$, the upper-right element of the U matrix has to vanish, and it becomes

$$\begin{pmatrix} 0 & 0 \\ 2\chi & 0 \end{pmatrix}. \tag{9.118}$$

As μ becomes larger from $\mu < \chi$ to $\mu > \chi$, the U matrix has to go through this triangular form. This is the same formalism that we saw in chapter 3.

From these three conditions we should be able to compute the exponential form of

$$e^{Uz}. \tag{9.119}$$

The usual problem is whether for the above quantity we can obtain an analytical expression. Let us proceed by writing a Taylor expansion, but we need to first calculate U^N. We can manage this calculation when $N = 2$. However, for an arbitrary large integer N, it is not a trivial problem. In this section this is the problem we would like to address.

If χ is greater than μ, we write U of equation (9.112) as

$$U = k\begin{pmatrix} e^{\zeta} & 0 \\ 0 & e^{-\zeta} \end{pmatrix}\begin{pmatrix} 0 & -1 \\ 1 & 0 \end{pmatrix}\begin{pmatrix} e^{-\zeta} & 0 \\ 0 & e^{\zeta} \end{pmatrix}, \tag{9.120}$$

with e^{ζ} given in equation (9.113). This is a similarity transformation of

$$\begin{pmatrix} 0 & -1 \\ 1 & 0 \end{pmatrix}, \tag{9.121}$$

with respect to a squeeze matrix

$$B = \begin{pmatrix} e^{\zeta} & 0 \\ 0 & e^{-\zeta} \end{pmatrix}. \tag{9.122}$$

The role of this squeeze matrix is quite different from that of equation (9.105). It does not depend on z.

Let us go back to the U matrix; we can write U^N as

$$k^N \begin{pmatrix} e^{\zeta} & 0 \\ 0 & e^{-\zeta} \end{pmatrix} \begin{pmatrix} 0 & -1 \\ 1 & 0 \end{pmatrix}^N \begin{pmatrix} e^{-\zeta} & 0 \\ 0 & e^{\zeta} \end{pmatrix} \tag{9.123}$$

and

$$\exp\left[kz\begin{pmatrix} 0 & -1 \\ 1 & 0 \end{pmatrix} \right] = \begin{pmatrix} \cos(kz) & -\sin(kz) \\ \sin(kz) & \cos(kz) \end{pmatrix}. \tag{9.124}$$

The exponential form e^{Uz} of equation (9.119) then becomes

$$\begin{pmatrix} e^{\zeta} & 0 \\ 0 & e^{-\zeta} \end{pmatrix} \begin{pmatrix} \cos(kz) & -\sin(kz) \\ \sin(kz) & \cos(kz) \end{pmatrix} \begin{pmatrix} e^{-\zeta} & 0 \\ 0 & e^{\zeta} \end{pmatrix} \tag{9.125}$$

and the transformation matrix of equation (9.107) takes the form

$$M(\chi, \mu, z) = e^{-\lambda z} \begin{pmatrix} \cos(kz) & -e^{2\zeta}\sin(kz) \\ e^{-2\zeta}\sin(kz) & \cos(kz) \end{pmatrix}, \tag{9.126}$$

with k and e^{ζ} given in equation (9.113).

If μ is greater than χ, the off-diagonal elements of equation (9.111) have the same sign and we can go through a similar calculation. The result is

$$M(\chi, \mu, z) = e^{-\lambda z} \begin{pmatrix} \cosh(kz) & e^{2\zeta}\sinh(kz) \\ e^{-2\zeta}\sinh(kz) & \cosh(kz) \end{pmatrix}, \tag{9.127}$$

with k and e^{ζ} given in equation (9.116).

If χ and μ are equal, the U matrix becomes

$$U = \begin{pmatrix} 0 & 0 \\ 2\chi & 0 \end{pmatrix}, \tag{9.128}$$

with the property

$$U^2 = \begin{pmatrix} 0 & 0 \\ 2\chi & 0 \end{pmatrix}^2 = 0, \tag{9.129}$$

and the transformation matrix becomes

$$\begin{pmatrix} 1 & 0 \\ 2\chi z & 1 \end{pmatrix}. \tag{9.130}$$

If we consider again the case with $\chi > \mu$, we see that the parameter μ gradually increases to a value greater than χ. This means that from equations (9.113) to (9.116) there is a singularity involved in the expression of $e^{2\zeta}$. We saw in chapter 3 that even though this is not an analytic continuity it is a tangential continuity.

In order to motivate the procedure followed above, let us consider what we should do if the matrix cannot be diagonalized by a unitary transformation. Starting from the U matrix of equation (9.111), if $\chi > \mu$, it was possible to bring U into the form

$$k\begin{pmatrix} 0 & -1 \\ 1 & 0 \end{pmatrix},$$ (9.131)

where the similarity transformation matrix of equation (9.122) is not unitary, but rather is a symmetric squeeze matrix. Additionally, the property

$$\begin{pmatrix} 0 & -1 \\ 1 & 0 \end{pmatrix}^2 = \begin{pmatrix} 1 & 0 \\ 0 & 1 \end{pmatrix} \quad \text{and} \quad \begin{pmatrix} 0 & -1 \\ 1 & 0 \end{pmatrix}^3 = \begin{pmatrix} 0 & -1 \\ 1 & 0 \end{pmatrix},$$ (9.132)

was used to deal with the Taylor expansion.

For $\mu > \chi$ we used

$$\begin{pmatrix} 0 & 1 \\ 1 & 0 \end{pmatrix}^2 = \begin{pmatrix} 1 & 0 \\ 0 & 1 \end{pmatrix} \quad \text{and} \quad \begin{pmatrix} 0 & 1 \\ 1 & 0 \end{pmatrix}^3 = \begin{pmatrix} 0 & 1 \\ 1 & 0 \end{pmatrix}.$$ (9.133)

If $\mu = \chi$ the U matrix becomes triangular and

$$\begin{pmatrix} 0 & 0 \\ 2\chi & 0 \end{pmatrix}^2 = 0.$$ (9.134)

The Taylor expansion truncates.

By using these properties of two-by-two matrices, we were able to deal with the problem even though not all of them can be diagonalized. The triangular matrix of equation (9.134) cannot be diagonalized. The matrix of equation (9.133) can be diagonalized with the diagonal elements of 1 and -1. The two-by-two matrix of equation (9.131) can also be diagonalized, but the eigenvalues are the imaginary numbers i and $-i$. However, imaginary numbers are not too convenient for computer mathematics. Thus, we had to resort to the method presented here.

9.3.2 Correspondence to space–time symmetries

Since the Lorentz group provides the basic mathematical framework for ray and polarization optics, let us see how optical activities can be translated into the four-by-four Lorentz transformation matrices applicable to the Minkowski space of (t, z, x, y). In this convention, the momentum–energy four-vector is $\left(E, p_z, p_x, p_y\right)$. Let us now translate the two-by-two matrices given in this section into the language of Wigner's little groups given in chapter 3.

If the particle is massive, there is a Lorentz frame in which it is at rest and this four-vector momentum is proportional to

$$P = (1, 0, 0, 0).$$ (9.135)

This vector is invariant under rotations around the z-axis. We can Lorentz boost the four-momentum of equation (9.135) along the z-direction using the Lorentz boost matrix

$$B = \begin{pmatrix} \cosh\zeta & \sinh\zeta & 0 & 0 \\ \sinh\zeta & \cosh\zeta & 0 & 0 \\ 0 & 0 & 1 & 0 \\ 0 & 0 & 0 & 1 \end{pmatrix}. \tag{9.136}$$

This Lorentz boost matrix corresponds to the squeeze matrix equation (9.68) [11] where $\zeta = \eta/2$. Now the four-momentum of equation (9.135) is invariant under the rotation matrix

$$R(\chi z) = \begin{pmatrix} 1 & 0 & 0 & 0 \\ 0 & \cos(\chi z) & -\sin(\chi z) & 0 \\ 0 & \sin(\chi z) & \cos(\chi z) & 0 \\ 0 & 0 & 0 & 1 \end{pmatrix}, \tag{9.137}$$

where $\chi = \theta/2$. Thus, the matrix

$$B(\zeta)R(\chi z)B(-\zeta) \tag{9.138}$$

leaves the four-momentum of equation (9.135) invariant.

While this matrix performs a rotation around the y-axis in the particle's rest frame, we can also rotate this four-momentum around the z-axis without changing it. This is what Wigner's little group is about for the particle with mass m. Although the matrix of equation (9.138) does not change the momentum, it rotates the spin direction of the particle in the rest frame. This is why the little group is not a trivial mathematical device. It is known that the rotation matrix of equation (9.137) corresponds to the rotation matrix of equation (9.98) [11]. Thus the two-by-two rotation matrix of equation (9.98), together with the squeeze matrix of equation (9.69), generates the little group for particles with non-zero mass.

If the particle has imaginary mass, we can start with the four-momentum proportional to

$$P = (E, p, 0, 0). \tag{9.139}$$

If E is smaller than p, it can be brought to the Lorentz frame where the four-vector becomes

$$P = (0, 1, 0, 0), \tag{9.140}$$

where $P^2 = -1$, a negative number. This means the particle mass is imaginary. The Lorentz boost matrix takes the same form as equation (9.136), with

$$\tanh(2\eta) = \frac{E}{p}. \tag{9.141}$$

The four-momentum of equation (9.140) is also invariant under the Lorentz boost

$$S(\mu z) = \begin{pmatrix} \cosh(2\mu z) & 0 & 0 & \sinh(2\mu z) \\ 0 & 1 & 0 & 0 \\ \sinh(2\mu z) & 0 & 0 & \cosh(2\mu z) \\ 0 & 0 & 0 & 1 \end{pmatrix} \tag{9.142}$$

along the x-direction. Here again the four-momentum of equation (9.140) is invariant under rotations around the z-axis.

The above four-by-four matrix corresponds to the two-by-two squeeze matrix of equation (9.104) applicable to optical activities [11]. Thus, this squeeze matrix, together with the squeeze matrix of equation (9.122), generate the little group for imaginary mass particles.

Finally we consider a massless particle with four-momentum proportional to

$$(1, 1, 0, 0). \tag{9.143}$$

It is invariant under the rotation around the z-axis. In addition, Wigner [22] observed that it is invariant under the transformation given in table 3.2:

$$\begin{pmatrix} 1+\chi^2 & -\chi^2 & 2\chi & 0 \\ \chi^2 & 1-\chi^2 & 2\chi & 0 \\ 2\chi & -2\chi & 1 & 0 \\ 0 & 0 & 0 & 1 \end{pmatrix}, \tag{9.144}$$

where 2χ has been used for γ. This four-by-four matrix has a controversial history [15, 16], but the bottom line is that it corresponds to the triangular matrix of equation (9.128), and the variable 2χ performs gauge transformations.

It is interesting to note that optical activities can act as computational devices for the internal space–time symmetries of elementary particles.

References

[1] Azzam R M A G and Bashara N M 1999 *Ellipsometry and Polarized Light* (Amsterdam: Elsevier)
[2] Başkal S and Kim Y S 2001 Shear representations of beam transfer matrices *Phys. Rev.* E **63** 056606
[3] Başkal S and Kim Y S 2003 Lens optics as an optical computer for group contractions *Phys. Rev.* E **67** 056601
[4] Başkal S and Kim Y S 2009 ABCD matrices as similarity transformations of Wigner matrices and periodic systems in optics *J. Opt. Soc. Am.* A **26** 2049–54
[5] Başkal S and Kim Y S 2010 One analytic form for four branches of the *ABCD* matrix *J. Mod. Opt.* **57** 1251–9
[6] Başkal S and Kim Y S 2013 Lorentz group in ray and polarization optics *Mathematical Optics: Classical, Quantum and Computational Methods* ed V Lakshminarayanan, M L Calvo and T Alieva (Boca Raton, FL: Taylor and Francis) pp 303–49
[7] Dragoman D 2010 Polarization optics analogy of quantum wavefunctions in graphene *J. Opt. Soc. Am.* B **27** 1325–31
[8] Georgieva E and Kim Y S 2001 Iwasawa effects in multilayer optics *Phys. Rev.* E **64** 26602
[9] Georgieva E and Kim Y S 2003 Slide-rule-like property of Wigner's little groups and cyclic *S* matrices for multilayer optics *Phys. Rev.* E **68** 026606
[10] Gerrard A and Burch J M 1994 *Introduction to Matrix Methods in Optics* (New York: Dover)

[11] Han D, Kim Y S, and Noz M E 1997 Stokes parameters as a Minkowskian four-vector *Phys. Rev.* E **56** 6065–76

[12] Han D, Kim Y S, and Noz M E 1999 Wigner rotations and Iwasawa decompositions in polarization optics *Phys. Rev.* E **60** 1036–41

[13] Haus H A 1984 *Waves and Fields in Optoelectronics* (*Prentice-Hall Series in Solid State Physical Electronics*) (Englewood Cliffs, NJ: Prentice-Hall)

[14] Hawkes J and Latimer I 1995 *Lasers: Theory and Practice* (*Prentice-Hall International Series in Optoelectronics*) (Upper Saddle River, NJ: Prentice Hall)

[15] Kim Y S and Noz M E 1986 *Theory and Applications of the Poincaré Group* (Dordrecht: Springer)

[16] Kim Y S and Wigner E P 1990 Space–time geometry of relativistic particles *J. Math. Phys.* **31** 55–60

[17] Lohmann A W 1993 Image rotation, Wigner rotation, and the fractional Fourier transform *J. Opt. Soc. Am.* A **10** 2181

[18] Monzón J J and Sánchez-Soto L L 1999 Fully relativistic-like formulation of multilayer optics *J. Opt. Soc. Am.* A **16** 2013–8

[19] Monzón J J and Sánchez-Soto L L 2000 Fresnel formulas as Lorentz transformations *J. Opt. Soc. Am.* A **17** 1475–81

[20] Saleh B E A and Carl Teich M 2007 *Fundamentals of Photonics* (*Wiley Series in Pure and Applied Optics*) 2nd edn (Hoboken, NJ: Wiley)

[21] Sánchez-Soto L L, Monzón J J, Barriuso A G, and Cariñena J F 2012 The transfer matrix: a geometrical perspective *Phys. Rep.* **513** 191–227

[22] Wigner E P 1939 On unitary representations of the inhomogeneous Lorentz group *Ann. Math.* **40** 149–204

[23] Yariv A 1989 *Quantum Electronics* 3rd edn (Hoboken, NJ: Wiley)

IOP Publishing

Mathematical Devices for Optical Sciences

Sibel Başkal, Young S Kim, and Marilyn E Noz

Chapter 10

Polarization optics

Although the two-by-two representation of the Lorentz group was introduced to physics as the language of space–time symmetries of relativistic particles [2, 31], it also provides an interesting set of mathematical tools which can also be used for studying polarization optics [6, 8, 23, 24, 26].

The standard language for optical polarizations is the Jones-matrix formalism [18, 25, 29], consisting of the Jones vectors, Mueller matrices, and Stokes parameters. These can be all formulated in terms of the six-parameter Lorentz group [11]. Other appropriate representations have also been proposed [4, 5, 9, 12, 24, 27, 30]. Among the many different representations of Lorentz group, it has been shown [11] that the bilinear conformal representation [2] is most convenient for polarization optics. Here we shall connect that representation with the Jones-matrix formalism.

It is usual to consider that polarized light is propagating along the z-direction. Therefore, the traditional approach is to consider the x- and y-components of the electric fields. It is their amplitude ratio and phase difference that determine the state of polarization. Thus, polarization can be changed either by adjusting the amplitudes or by changing the relative phases, or both. The polarization of light can be described as two-component column vectors, and the two-by-two matrices of the Lorentz group (see section 2.2) applicable to those vectors.

Phase-shift filters and attenuation can be represented, respectively, by the three-parameter rotation group and the three-parameter subgroup of the Lorentz group with real matrices $Sp(2)$. It is shown that the rotation group is the underlying language for optical filters causing phase shifts between the two transverse components [11]. It is known that this $Sp(2)$ group is the proper language for attenuation filters [11, 23]. The Lorentz group has another three-parameter subgroup which is like the two-dimensional Euclidean group $E(2)$. This $E(2)$-like group has been applied to optics [22], in particular with respect to possible optical filters. These three-parameter subgroups of the Lorentz group were described in chapter 2.

doi:10.1088/2053-2563/aafe78ch10

We discuss first in section 10.1 the Jones vector as the traditional language for studying the two-component light vector and show that the Jones-matrix formalism is indeed a representation of the Lorentz group. It is shown that the bilinear conformal representation of the Lorentz group is the natural scientific language for polarization optics. We also discuss the squeeze and phase shift, the rotation of the polarization axes, and the combined effects of squeeze and attenuation. It is shown that the combined effect of attenuation and phase-shift filters leads to a six-parameter two-by-two matrix and that the attenuation and phase-shift filters have their own sub-representations.

In section 10.2 the mathematics of the former section are translated into matrices corresponding to optical filters. It is shown that one of those sub-representations leads to a new kind of optical filter having the symmetry of the $E(2)$-like subgroup of the Lorentz group. We also discuss possible applications of these new filters.

The polarization plane is not always perpendicular to the propagation direction of the light wave. The Jones-matrix formalism can also be extended to the case where the polarization plane is not perpendicular to the direction of propagation systems within the framework of the Lorentz-group representation. In section 10.3 we show that the formalism developed here can accommodate the cases where the polarization coordinates are squeezed or sheared. It is known that one of the $E(2)$-like transformations leads to a *shear* transformation.

10.1 Jones vector, phase shifters, and attenuators

As covered in standard optics textbooks [17, 28], the traditional language for studying the two-component light vector is the Jones-matrix formalism. For a plane wave propagating along the z-direction, the polarizations are along the x- and y-directions. Then the electric field vector can be written as

$$
\begin{aligned}
E_x &= A \cos (kz - \omega t + \phi_1), \\
E_y &= B \cos (kz - \omega t + \phi_2).
\end{aligned}
\tag{10.1}
$$

In this equation, A and B are real and positive numbers representing the amplitudes. The angles ϕ_1 and ϕ_2 represent the phases of the x- and y-components. This form is useful in classical optics as well as being applicable to coherent and squeezed states of light [7, 21].

When using this formalism, the two transverse components of the electric field are generally combined into one column matrix

$$
\begin{pmatrix} E_x \\ E_y \end{pmatrix} = \begin{pmatrix} A \exp \{i(kz - \omega t + \phi_1)\} \\ B \exp \{i(kz - \omega t + \phi_2)\} \end{pmatrix}.
\tag{10.2}
$$

Here the exponential form is used for the sinusoidal function. This column matrix is known as the Jones vector [18, 19].

The polarization content can be determined by the ratio

$$\frac{E_y}{E_x} = \left(\frac{B}{A}\right) e^{i(\phi_2 - \phi_1)}. \tag{10.3}$$

This can in turn be written as one complex number:

$$w = re^{i\phi}. \tag{10.4}$$

In this equation

$$r = \frac{B}{A} \quad \text{and} \quad \phi = \phi_2 - \phi_1.$$

These two real numbers measure the degree of polarization, and they are the amplitude ratio and the phase difference, respectively. The transformation properties of this complex number w were discussed in section 2.7. It is only when the light beam goes through an optical filter whose transmission properties are not isotropic that the transformation takes place.

The Jones-matrix formalism [25] that can be found in most of the existing textbooks [16, 17], starts with the projection operator

$$\begin{pmatrix} 1 & 0 \\ 0 & 0 \end{pmatrix}. \tag{10.5}$$

This projection operator is applicable to the Jones vector of equation (10.2). It can be easily seen that this operator keeps the x electric field component and completely eliminates the y-component. This does not consistently describe the real world if the attenuation factor in the y-direction is much greater than that in the x-direction. It is therefore desirable to replace this projection operator by an attenuation matrix which matches the real world more closely. Since there are two perpendicular transverse directions, it is possible that the attenuation coefficient along one transverse direction could be different from that along the other direction. Hence, there exists the *polarization coordinate* in which the attenuation can be described as follows [5, 12, 13, 24]:

$$\begin{pmatrix} e^{-\eta_1} & 0 \\ 0 & e^{-\eta_2} \end{pmatrix} = e^{-(\eta_1 + \eta_2)/2} \begin{pmatrix} e^{\eta/2} & 0 \\ 0 & e^{-\eta/2} \end{pmatrix}. \tag{10.6}$$

Here we have used $\eta = \eta_2 - \eta_1$. The attenuation matrix in equation (10.6) becomes the projection matrix if η_1 is very close to zero and η_2 becomes infinitely large. The projection operator of equation (10.5) is thus seen to be a special case of the attenuation matrix in equation (10.6). Since the exponential factor $e^{-(\eta_1 + \eta_2)/2}$ reduces both components at the same rate it does not affect the state of polarization. Therefore, the polarization effect is determined only by the squeeze matrix

$$B(\eta) = \begin{pmatrix} e^{\eta/2} & 0 \\ 0 & e^{-\eta/2} \end{pmatrix}. \tag{10.7}$$

This type of mathematical operation was seen in both the squeezed state of light and the Lorentz boosts of spinors. We are thus replacing the projection operator of equation (10.5) by the squeeze matrix.

Furthermore, the optical filter with two different values of the index of refraction along the two orthogonal directions provides another basic element. We can write the effect of this filter as

$$\begin{pmatrix} e^{-i\delta_1} & 0 \\ 0 & e^{-i\delta_2} \end{pmatrix} = e^{-i(\delta_1+\delta_2)/2} \begin{pmatrix} e^{-i\delta/2} & 0 \\ 0 & e^{i\delta/2} \end{pmatrix}. \tag{10.8}$$

Here we have used $\delta = \delta_1 - \delta_2$. In measurement processes, the overall phase factor $e^{-i(\delta_1+\delta_2)/2}$ cannot be detected, and can therefore be deleted. The polarization effect of the filter is solely determined by the matrix

$$P(\delta) = \begin{pmatrix} e^{-i\delta/2} & 0 \\ 0 & e^{i\delta/2} \end{pmatrix}. \tag{10.9}$$

This matrix, which we shall call a phase shifter, appears like a rotation matrix around the z-axis in the theory of rotation groups, but it plays a completely different role here.

Furthermore, a rotation matrix is needed

$$R(\theta) = \begin{pmatrix} \cos(\theta/2) & -\sin(\theta/2) \\ \sin(\theta/2) & \cos(\theta/2) \end{pmatrix} \tag{10.10}$$

since the x- and y-axes are not always the polarization axes. This matrix is also familiar to us from the Lorentz group. The systematic combination of the three components given in equations (10.7), (10.9), and (10.10) constitute the traditional Jones-matrix formalism.

10.1.1 Squeeze and phase shift

While the effect of the phase-shift matrix $P(\delta)$ of equation (10.9) on the Jones vector is well known, the effect of the squeeze matrix of equation (10.7) is not as well addressed in the literature. Here we will discuss the combined effect of these two matrices. We note that they are both diagonal and they commute.

If the squeeze matrix of equation (10.7) is applied to the Jones vector, the result is

$$\begin{pmatrix} e^{\eta/2} & 0 \\ 0 & e^{-\eta/2} \end{pmatrix} \begin{pmatrix} E_x \\ E_y \end{pmatrix} = \begin{pmatrix} e^{\eta/2} E_x \\ e^{-\eta/2} E_y \end{pmatrix}. \tag{10.11}$$

It is thus apparent that the squeeze transformation expands one amplitude, while contracting the other. This results in the product of the amplitude remaining invariant. We illustrate this squeeze transformation in figure 10.1.

To gain a better understanding of phase shifts, it is possible to write the Jones vector in the form

$$\begin{pmatrix} \exp(ikz) \\ \exp[i(kz - \pi/2)] \end{pmatrix}. \tag{10.12}$$

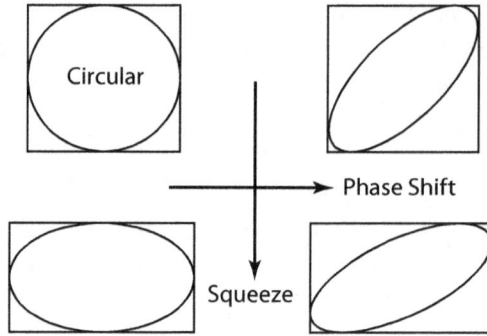

Figure 10.1. Squeeze and phase shift. Even though the squeeze and phase shifts are both elliptic deformations, they are performed differently [4].

Then the real part of this equation is

$$\text{Re}\begin{pmatrix} x \\ y \end{pmatrix} = \begin{pmatrix} \cos(kz) \\ \sin(kz) \end{pmatrix}. \tag{10.13}$$

This corresponds to a circular polarization with

$$x^2 + y^2 = 1. \tag{10.14}$$

If the phase-shift matrix is applied, we obtain the vector

$$\begin{pmatrix} x \\ y \end{pmatrix} = \begin{pmatrix} \cos(kz + \delta/2) \\ \sin(kz - \delta/2) \end{pmatrix}. \tag{10.15}$$

This vector can be written as

$$\begin{pmatrix} x \\ y \end{pmatrix} = \begin{pmatrix} \cos(kz - \pi/4 + \alpha) \\ \cos(kz - \pi/4 - \alpha) \end{pmatrix}, \tag{10.16}$$

with

$$\alpha = \frac{\delta}{2} + \frac{\pi}{4}. \tag{10.17}$$

Then we obtain

$$\begin{aligned} x + y &= 2(\cos \alpha)\cos(kz - \pi/4), \\ x - y &= -2(\sin \alpha)\sin(kz - \pi/4), \end{aligned} \tag{10.18}$$

and

$$\frac{(x + y)^2}{4(\cos \alpha)^2} + \frac{(x - y)^2}{4(\sin \alpha)^2} = 1. \tag{10.19}$$

This results in an elliptic polarization.

This simple squeeze operation of equation (10.7) changes the amplitudes, and commutes with the phase-shift matrix. The combined effect is illustrated in figure 10.1.

10.1.2 Rotation of the polarization axes and combined effects

When the electric field components take the form of equation (10.13), and the polarization coordinate is the same as the xy-coordinate, then we can apply the phase shifter in equation (10.9) directly to the column matrix of equation (10.2). If we then rotate the polarization coordinate by an angle $\theta/2$, or by using the matrix

$$R(\theta) = \begin{pmatrix} \cos(\theta/2) & -\sin(\theta/2) \\ \sin(\theta/2) & \cos(\theta/2) \end{pmatrix},$$
(10.20)

the phase shifter then has the form

$$\begin{aligned} P(\theta, \delta) &= R(\theta)P(\delta)R(-\theta) \\ &= \begin{pmatrix} \cos(\delta/2) + i\sin(\delta/2)\cos\theta & i\sin(\delta/2)\sin\theta \\ i\sin(\delta/2)\sin\theta & \cos(\delta/2) - i\sin(\delta/2)\cos\theta \end{pmatrix}. \end{aligned}$$
(10.21)

If we then rotate the polarization coordinate system by 45°, the phase shifter matrix has the form

$$Q(\delta) = \begin{pmatrix} \cos(\delta/2) & i\sin(\delta/2) \\ i\sin(\delta/2) & \cos(\delta/2) \end{pmatrix}.$$
(10.22)

To illustrate the effect this matrix has on the polarized beams, we can start with the circularly polarized wave

$$\begin{pmatrix} x \\ y \end{pmatrix} = \begin{pmatrix} 1 \\ -i \end{pmatrix} e^{(ikz - i\omega t)}.$$
(10.23)

The real part of this equation is then

$$\mathrm{Re}\begin{pmatrix} x \\ y \end{pmatrix} = \begin{pmatrix} \cos(kz - \omega t) \\ \sin(kz - \omega t) \end{pmatrix}.$$
(10.24)

Hence we obtain the familiar equation for the circle

$$x^2 + y^2 = 1.$$
(10.25)

When the phase shifter of equation (10.22) is applied to the Jones vector of equation (10.23), we obtain the real part as

$$\begin{pmatrix} [\cos(\delta/2) + \sin(\delta/2)]\cos(kz - \omega t) \\ [\cos(\delta/2) - \sin(\delta/2)]\sin(kz - \omega t) \end{pmatrix}.$$
(10.26)

Using

$$\begin{aligned} \cos(\delta/2) &= \cos([\delta/2 + \pi/4] - \pi/4), \\ -\sin(\delta/2) &= \cos([\delta/2 + \pi/4] + \pi/4), \end{aligned}$$
(10.27)

we can then write

$$\cos(\delta/2) - \sin(\delta/2) = \sqrt{2}\cos(\delta/2 + \pi/4),$$
$$\cos(\delta/2) + \sin(\delta/2) = \sqrt{2}\sin(\delta/2 + \pi/4).$$
(10.28)

Hence after the phase shift is applied, the Jones vector becomes

$$\begin{pmatrix} [\sqrt{2}\sin\alpha]\cos(kz - \omega t) \\ [\sqrt{2}\cos\alpha]\sin(kz - \omega t) \end{pmatrix},$$
(10.29)

where

$$\alpha = \frac{\delta}{2} + \frac{\pi}{4}.$$
(10.30)

The x- and y-components then satisfy the equation

$$\frac{x^2}{(\sqrt{2}\sin\alpha)^2} + \frac{y^2}{(\sqrt{2}\cos\alpha)^2} = 1,$$
(10.31)

which is an elliptic polarization. These steps are illustrated in figure 10.2.

If we now consider rotations of the squeeze matrix.

$$B(\theta, \eta) = R(\theta)B(\eta)R(-\theta),$$
(10.32)

we can obtain

$$B(\theta, \eta) = \begin{pmatrix} \cosh(\eta/2) + \sinh(\eta/2)\cos\theta & \sinh(\eta/2)\sin\theta \\ \sinh(\eta/2)\sin\theta & \cosh(\eta/2) - \sinh(\eta/2)\cos\theta \end{pmatrix}.$$
(10.33)

We are familiar with this squeeze operation from section 9.1.2. This changes the amplitudes. If the squeeze angle becomes 45°, we use the notation $S(\eta)$ for this special case of $B(\theta, \eta)$, where it takes the form

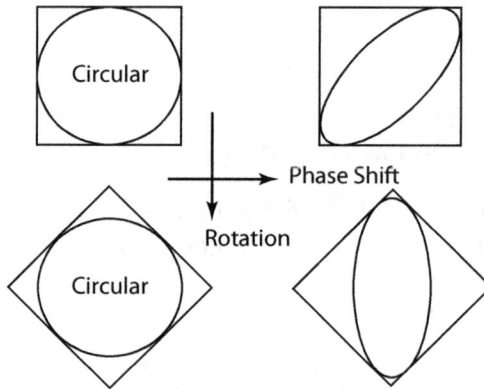

Figure 10.2. Phase shift and rotation. There is a 45° rotation [4].

$$S(\eta) = \begin{pmatrix} \cosh(\eta/2) & \sinh(\eta/2) \\ \sinh(\eta/2) & \cosh(\eta/2) \end{pmatrix}. \tag{10.34}$$

The question now arises as to what happens when two squeeze transformations are performed in two different directions. The multiplication of two squeeze matrices does not just lead to another squeeze matrix but it leads to a squeeze matrix preceded by a rotation matrix. This can be expressed as [3]

$$B(\theta, \lambda)B(0, \eta) = B(\phi, \xi)R(\omega), \tag{10.35}$$

where $R(\omega)$ is a rotation around the z-axis by angle ω. This aspect of the squeeze operation is well known from the squeezed state of light, and has been discussed extensively in the literature [6, 21, 23]. The relation between the parameters of equation (10.35) is found to be

$$\cosh \xi = \cosh \eta \, \cosh \lambda + \sinh \eta \, \sinh \lambda \, \cos \theta,$$
$$\tan \phi = \frac{\sin \theta[\sinh \lambda + \tanh \eta(\cosh \lambda - 1)\cos \theta]}{\sinh \lambda \cos \theta + \tanh \eta[1 + (\cosh \lambda - 1)\cos^2 \theta]}, \tag{10.36}$$
$$\tan \omega = \frac{2(\sin \theta)[\sinh \lambda \sinh \eta + C_-\cos \theta]}{C_+ + C_-\cos(2\theta) + 2 \sinh \lambda \sinh \eta \cos \theta},$$

where

$$C_\pm = (\cosh \lambda \pm 1)(\cosh \eta \pm 1). \tag{10.37}$$

Indeed, we can write equation (10.35) as

$$R(\omega) = B(\phi, -\xi) \, B(\theta, \lambda) \, B(0, \eta), \tag{10.38}$$

meaning that three squeeze transformations lead to one rotation. This rotation is known as the Wigner rotation in the literature [15, 21].

If the angle θ is 90°, the calculation becomes simpler, and

$$B(\lambda)B(\eta) = B(\phi, \xi)R(\omega), \tag{10.39}$$

where

$$\cosh \xi = \cosh \eta \, \cosh \lambda,$$
$$\tan \phi = \frac{\sinh \lambda}{\tanh \eta}, \tag{10.40}$$
$$\tan \omega = \frac{\sinh \lambda \sinh \eta}{\cosh \eta + \cosh \lambda}.$$

In these calculations we used the kinematics of Lorentz transformations. While it does not appear possible to perform experiments using high-energy particles, it is gratifying to note that this experiment is possible in polarization optics.

10.1.3 The $SL(2, c)$ content of polarization optics

The phase shifter of equation (10.9) can be written as

$$P(\delta) = \exp(-i\delta J_1), \quad \text{with} \quad J_1 = \frac{1}{2}\begin{pmatrix} 0 & 1 \\ 1 & 0 \end{pmatrix}. \tag{10.41}$$

The rotation operator of equation (10.10) takes the form

$$R(\theta) = \exp(-i\theta J_2), \quad \text{with} \quad J_2 = \frac{1}{2}\begin{pmatrix} 0 & -i \\ i & 0 \end{pmatrix}. \tag{10.42}$$

The squeeze operator of equation (10.7) can also be written in the exponential form

$$B(\eta) = \exp(-i\eta K_3), \quad \text{with} \quad K_3 = \frac{i}{2}\begin{pmatrix} 1 & 0 \\ 0 & -1 \end{pmatrix}. \tag{10.43}$$

The generators J_1, J_2, and K_3 do not by themselves form a closed set of commutation relations, but rather their commutation relations produce three more generators which results in the full set of generators for the Lie algebra of the Lorentz group. The commutation relations for these six generators were given in equation (2.20).

The generators K_3 and J_2, however, are needed for the squeeze matrix of equation (10.34). The repeated application of these two generators to the attenuation matrix leads to the commutation relation

$$[J_2, K_3] = iK_1. \tag{10.44}$$

Indeed, as discussed previously in section 2.5, J_2, K_1, and K_3 form a closed set of commutation relations for the $Sp(2)$ group which is a subgroup of $SL(2, c)$ [23]. This three-parameter group has been extensively discussed in connection with squeezed states of light [21, 32].

If we consider only the phase shifters, the mathematics is basically repeated applications of J_1 and J_2, resulting in applications also of J_3, for which explicit two-by-two matrix forms are given in table 2.1. Thus, the phase-shift filters form an $SU(2)$- or $O(3)$-like group which is again a subgroup of $SL(2, c)$.

If we use both the attenuators and phase shifters, the result is the full $SL(2, c)$ group with six parameters. The transformation matrix, which we previously saw in chapter 2, is written as

$$G = \begin{pmatrix} \alpha & \beta \\ \gamma & \delta \end{pmatrix}, \tag{10.45}$$

with the condition that the determinant of G be 1: $\alpha\delta - \gamma\beta = 1$. The repeated application of two matrices of this kind results in

$$\begin{pmatrix} \alpha_2 & \beta_2 \\ \gamma_2 & \delta_2 \end{pmatrix}\begin{pmatrix} \alpha_1 & \beta_1 \\ \gamma_1 & \delta_1 \end{pmatrix}$$
$$= \begin{pmatrix} \alpha_2\alpha_1 + \beta_2\gamma_1 & \alpha_2\beta_1 + \beta_2\delta_1 \\ \gamma_2\alpha_1 + \delta_2\gamma_1 & \gamma_2\beta_1 + \delta_2\delta_1 \end{pmatrix}. \tag{10.46}$$

The most general form of the polarization transformation is the application of this algebra to the column matrix of equation (10.2).

The generators of the rotation and the phase shifters are Hermitian. Thus, they form a unitary subset of the G matrices of equation (10.45). Repeated applications of the phase shifters represents the three-parameter rotation-like subgroup. The generators J_2, K_1, and K_3 are all imaginary and they generate the real $Sp(2)$ group. Thus, the real subset of the G matrices represents the attenuation filters and their repeated applications.

10.2 New filters and possible applications

We should note at this point that the Lorentz group has another set of three-parameter subgroups. They are like the two-dimensional Euclidean group, $E(2)$. Let us consider one of them, defined in section 2.5 which is generated by the matrices J_3, N_1, and N_2, with

$$N_1 = K_1 - J_2 \qquad \text{and} \qquad N_2 = K_2 + J_1, \tag{10.47}$$

where

$$N_1 = \begin{pmatrix} 0 & i \\ 0 & 0 \end{pmatrix} \quad \text{and} \quad N_2 = \begin{pmatrix} 0 & 1 \\ 0 & 0 \end{pmatrix}. \tag{10.48}$$

These matrices satisfy a closed set of commutation relations:

$$[J_3, N_1] = iN_2, \quad [J_3, N_2] = -iN_1, \quad \text{and} \quad [N_1, N_2] = 0. \tag{10.49}$$

As shown in sections 2.5 and 3.1.2, these commutation relations are like those for the two-dimensional Euclidean group consisting of two translations and one rotation around the origin. This group has been studied extensively in connection with the space–time symmetries of massless particles, where J_3 and the two N generators correspond to the helicity and gauge degrees of freedom, respectively [14].

The physics of J_3 is well known through the phase shifter given in equation (10.9). If the angle δ is $\pi/2$, the phase shifter becomes a quarter-wave shifter, which we write as

$$Q = P(\pi/2) = \begin{pmatrix} e^{-i\pi/4} & 0 \\ 0 & e^{i\pi/4} \end{pmatrix}. \tag{10.50}$$

Then J_2 and K_1 are the quarter-wave conjugates of J_1 and K_2, respectively:

$$J_2 = QJ_1Q^{-1} \qquad \text{and} \qquad K_1 = -QK_2Q^{-1}. \tag{10.51}$$

Consequently,

$$N_2 = QN_1Q^{-1}. \tag{10.52}$$

The N generators lead to the following transformation matrices:

$$T_1(\tau) = \exp(-i\tau N_1) = \begin{pmatrix} 1 & \tau \\ 0 & 1 \end{pmatrix},$$

$$T_2(\tau) = \exp(-i\tau N_2) = \begin{pmatrix} 1 & -i\tau \\ 0 & 1 \end{pmatrix}. \tag{10.53}$$

It is clear that T_1 is the quarter-wave conjugate of T_2. We can now concentrate on the transformation matrix T_1.

If T_1 is applied to the column matrix of equation (10.2),

$$\begin{pmatrix} 1 & \tau \\ 0 & 1 \end{pmatrix} \begin{pmatrix} E_x \\ E_y \end{pmatrix} = \begin{pmatrix} E_x + \tau E_y \\ E_y \end{pmatrix}. \tag{10.54}$$

This new filter superposes the y-component of the electric field to the x-component with an appropriate constant, but it leaves the y-component invariant.

Let us examine how this is achieved. The generator N_1 consists of J_2 which generates rotations around the y-axis, and K_1 which generates a squeeze along the 45° axis. Physically, J_2 generates optical activities. Thus, the new filter consists of a suitable combination of these two operations. In both cases, we have to take into account the overall attenuation factor. This can be measured by the attenuation of the y-component, which is not affected by the symmetry operation of equation (10.54).

Is it possible to produce optical filters of this kind? Starting from an optically active material, we can introduce an asymmetry in attenuation to it by either mechanical or electrical means. Another approach would be to alternately pile up J_2-type and K_1-type layers. In either case, it is interesting to note that the combination of these two effects produces a special effect predicted from the Lorentz group.

The $E(2)$-like symmetry includes transformations generated by J_1. This matrix is given in equation (10.9) and the physics associated to it is well understood. Let us apply this matrix to T_1 from left and from right. Then

$$P(\delta)T_1(\tau) = \begin{pmatrix} e^{-i\delta/2} & e^{-i\delta/2}\tau \\ 0 & e^{i\delta/2} \end{pmatrix},$$

$$T_1(\tau)P(\delta) = \begin{pmatrix} e^{-i\delta/2} & e^{i\delta/2}\tau \\ 0 & e^{i\delta/2} \end{pmatrix}. \tag{10.55}$$

This leads to

$$P(-\delta)T_1(\tau)P(\delta) = T_1(\delta, \tau)$$
$$= \begin{pmatrix} 1 & e^{i\delta}\tau \\ 0 & 1 \end{pmatrix}. \tag{10.56}$$

We can of course obtain the $P(-\delta)$ filter by rotating the $P(\delta)$ around the z-axis by 90°. It is thus possible to add a phase factor to the τ variable using phase shifters of the type $P(\delta)$.

The three-dimensional rotation group occupies an important place in many different branches of physics. The group $Sp(2)$ also is useful in a number of fields including optics [1, 10, 20, 21]. Thus, traditional attenuation and phase-shift filters may be useful in constructing analog computers performing the symmetry operations of these groups.

The group $E(2)$ is somewhat new in optics [22]. However, as was shown in chapter 3 it deals with translations and rotations on a flat surface. This filter may therefore be useful as a computational device for recording and reading two-dimensional maps. Let us rewrite the T_2 matrix of equation (10.53) as

$$\begin{pmatrix} 1 & \beta \\ 0 & 1 \end{pmatrix}. \tag{10.57}$$

This matrix has an interesting algebraic property:

$$\begin{pmatrix} 1 & \beta_1 \\ 0 & 1 \end{pmatrix}\begin{pmatrix} 1 & \beta_2 \\ 0 & 1 \end{pmatrix} = \begin{pmatrix} 1 & \beta_1 + \beta_2 \\ 0 & 1 \end{pmatrix}. \tag{10.58}$$

The matrix can therefore be used for converting multiplication into addition, like the logarithmic function. This is one of the most basic operations in computational machines.

As for a more immediate application, let us consider lens optics. It is a trivial laboratory operation to rotate a given filter around the z-axis by 90°. If we rotate the matrix of equation (10.57), the result is the matrix of the form

$$\begin{pmatrix} 1 & 0 \\ \beta & 1 \end{pmatrix}. \tag{10.59}$$

This form together with the original form of equation (10.57) serve as lens and translation matrices, respectively, in para-axial optics as we have seen in section 9.1.3. Indeed, a system of polarization filters can serve as an analog computer for a multi-lens system.

Furthermore, the matrix of the form given in equation (10.57) represents a *shear* transformation. This is one of the basic deformations in engineering applications.

The Lorentz group was introduced to physics as the basic language for space–time symmetries of elementary particles [31], but it is becoming increasingly prominent in many branches of physics and engineering including classical and quantum optics. The optical filters may provide excellent calculational tools for the Lorentz group. Thus, these filters may be useful as components of future computers.

10.3 Non-orthogonal coordinate systems

The rotation and squeeze transformations discussed section 10.1 are directly applicable to coordinate transformations. Thus, they are applicable to the case of where the polarization occurs along a skew or squeezed coordinate system. The formalism is therefore also applicable to the case where the polarization coordinate is sheared.

Since light polarization is caused by anisotropic crystals, the polarization coordinate is not always orthogonal. Let us consider first the case where the light polarization is along a pair of skewed or squeezed axes.

It was noted here that the $Sp(2)$-like subgroup can take care of attenuation filters. A transformation matrix of this subgroup can also transform the orthogonal coordinate system into a squeezed coordinate system. The matrix takes the form

$$\begin{pmatrix} x' \\ y' \end{pmatrix} = \begin{pmatrix} \cosh(\eta/2) & \sinh(\eta/2) \\ \sinh(\eta/2) & \cosh(\eta/2) \end{pmatrix} \begin{pmatrix} x \\ y \end{pmatrix}. \tag{10.60}$$

The idea is to transform the squeezed coordinate system into the orthogonal system using the matrix or its inverse given in the above expression. Next, we can perform the polarization algebra developed here for the orthogonal coordinate system. We then can transform the result obtained in the orthogonal system back to the original squeezed coordinate system.

It is interesting to note that the transformation matrix given in equation (10.60) is one of the transformation matrices within the framework of the Lorentz-group representation developed in this book, and there is no need to make the existing mathematics more complicated. The story is the same for the shear transformation which takes the form

$$\begin{pmatrix} x' \\ y' \end{pmatrix} = \begin{pmatrix} 1 & b \\ 0 & 1 \end{pmatrix} \begin{pmatrix} x \\ y \end{pmatrix}. \tag{10.61}$$

This transformation is also well within the framework discussed here.

The polarization plane is not always perpendicular to the direction of the propagation. We can take care of this problem by extending our two-by-two formalism of equation (10.45) into the three-by-three form

$$\begin{pmatrix} \alpha & \beta & 0 \\ \gamma & \delta & 0 \\ 0 & 0 & 1 \end{pmatrix} \tag{10.62}$$

applicable to the (z, x, y) coordinate system. The polarization plane can be rotated around the y-axis by

$$\begin{pmatrix} 1 & 0 & 0 \\ 0 & \cos\theta/2 & -\sin\theta/2 \\ 0 & \sin\theta/2 & \cos\theta/2 \end{pmatrix}. \tag{10.63}$$

Using this matrix or its inverse, we can bring the problem to the orthogonal coordinate system. After doing the standard polarization algebra, we can go back to the original coordinate system.

References

[1] Abraham R and Marsden J E 2008 *Foundations of Mechanics* 2nd edn (Providence, RI: American Mathematical Society)

[2] Bargmann V 1947 Irreducible unitary representations of the Lorentz group *Ann. Math.* **48** 568

[3] Başkal S and Kim Y S 2005 Rotations associated with Lorentz boosts *J. Phys. A: Math. Gen.* **38** 6545–56

[4] Başkal S and Kim Y S 2013 Lorentz group in ray and polarization optics *Mathematical Optics: Classical, Quantum and Computational Methods* ed V Lakshminarayanan, M L Calvo and T Alieva (Boca Raton, FL: Taylor and Francis) pp 303–49

[5] Ben-Aryeh Y 2005 Nonunitary squeezing and biorthogonal scalar products in polarization optics *J. Opt.* B **7** S452–7

[6] Chiao R Y and Jordan T F 1988 Lorentz-group Berry phases in squeezed light *Phys. Lett.* A **132** 77–81

[7] Chirkin A S, Parashchuk D Y, and Orlov A A 1993 Quantum theory of two-mode interactions in optically anisotropic media with cubic nonlinearities: Generation of quadrature- and polarization-squeezed light *Quant. Electron.* **23** 870–4

[8] Cloude S R 1986 Group theory and polarization algebra *Optik* **75** 26–36

[9] Franssens G R 2015 Relativistic kinematics formulation of the polarization effects of Jones–Mueller matrices *J. Opt. Soc. Am.* A **32** 164–72

[10] Guillemin V and Sternberg S 2001 *Symplectic Techniques in Physics* (Cambridge: Cambridge University Press)

[11] Han D, Kim Y S, and Noz M E 1996 Polarization optics and bilinear representation of the Lorentz group *Phys. Lett.* A **219** 26–32

[12] Han D, Kim Y S, and Noz M E 1997 Jones-matrix formalism as a representation of the Lorentz group *J. Opt. Soc. Am.* A **14** 2290–8

[13] Han D, Kim Y S, and Noz M E 1997 Stokes parameters as a Minkowskian four-vector *Phys. Rev.* E **56** 6065–76

[14] Han D, Kim Y S, and Son D 1982 *E*(2)-like little group for massless particles and neutrino polarization as a consequence of gauge invariance *Phys. Rev.* D **26** 3717–25

[15] Han D, Kim Y S, and Son D 1987 Thomas precession, Wigner rotations, and gauge transformations *Class. Quant. Grav.* **4** 1777–83

[16] Hecht E 1970 Note on an operational definition of the Stokes parameters *Am. J. Phys.* **38** 1156–8

[17] Hecht E 2002 *Optics* 4th edn (Reading, MA: Addison-Wesley)

[18] Jones R C 1941 A new calculus for the treatment of optical systems I description and discussion of the calculus *J. Opt. Soc. Am.* **31** 488–93

[19] Jones R C 1947 A new calculus for the treatment of optical systems V. A more general formulation, and description of another calculus *J. Opt. Soc. Am.* **37** 107–10

[20] Kim Y S and Noz M E 1986 *Theory and Applications of the Poincaré Group* (Dordrecht: Springer)

[21] Kim Y S and Noz M E 1991 *Phase Space Picture of Quantum Mechanics: Group Theoretical Approach* (*Lecture Notes in Physics* vol 40) (Singapore: World Scientific)

[22] Kim Y S and Yeh L 1992 *E*(2)-symmetric two-mode sheared states *J. Math. Phys.* **33** 1237–46

[23] Kitano M and Yabuzaki T 1989 Observation of Lorentz-group Berry phases in polarization optics *Phys. Lett.* A **142** 321–5

[24] Opatrný T and Peřina J 1993 Non-image-forming polarization optical devices and Lorentz transformations an analogy *Phys. Lett.* A **181** 199–202

[25] Pedrotti F L and Pedrotti L S 1993 *Introduction to Optics* 2nd edn (Englewood Cliffs, NJ: Prentice Hall)

[26] Pellat-Finet P and Buasset M 1992 What is common to both polarization optics and relativistic kinematics? *Optik* **90** 101–6

[27] Red'kov V M 2011 Lorentz group theory and polarization of the light *Adv. Appl. Clifford Algebr.* **21** 203–20

[28] Saleh B E A and Carl Teich M 2007 *Fundamentals of Photonics* (*Wiley Series in Pure and Applied Optics*) 2nd edn (Hoboken, NJ: Wiley)

[29] Swindell W (ed) 1975 *Polarized Light* (Stroudsburg, PA: Dowden, Hutchinson, and Ross)

[30] Tudor T 2015 On a quasi-relativistic formula in polarization theory *Opt. Lett.* **40** 693

[31] Wigner E P 1939 On unitary representations of the inhomogeneous Lorentz group *Ann. Math.* **40** 149–204

[32] Yuen H P 1976 Two-photon coherent states of the radiation field *Phys. Rev.* A **13** 2226–43

IOP Publishing

Mathematical Devices for Optical Sciences

Sibel Başkal, Young S Kim, and Marilyn E Noz

Chapter 11

Stokes parameters and Poincaré sphere

Even before the work of Einstein and Minkowski, Henri Poincaré formulated the mathematics of Lorentz transformations, known as the Poincaré group. With the intent of analyzing the polarization of light, Poincaré also constructed a graphic illustration known as the Poincaré sphere [1, 6, 7]. We will see that the Poincaré sphere can be formulated in the language of special relativity. We note that the Poincaré sphere can also be formulated in terms of two-by-two matrices which in the later instance consists of four Stokes parameters. Hence we can show that the Poincaré sphere shares the same symmetry property as that of the Lorentz group, particularly in approaching Wigner's little groups. This will allow us to study the Lorentz symmetries of elementary particles from what we observe in optical laboratories.

The two-by-two formalism of the group of Lorentz transformations will be used for transformations of the coherency matrix in polarization optics. This matrix consists of four Stokes parameters and is like a Minkowskian four-vector under four-by-four Lorentz transformations. As the Poincaré sphere needs only three parameters and yet there are four Stokes parameters, we need to accommodate this fourth parameter. To do this, we note that the radius of the Poincaré sphere should be allowed to vary from its maximum value to its minimum, corresponding to the fully and minimal coherent polarization. As with the particle mass, the decoherence parameter in the Stokes formalism is invariant under Lorentz transformations. However, the Poincaré sphere, with a variable radius, provides the mechanism for the variations of the decoherence parameter. This variation gives a physical process whose mathematics is equivalent to a massless particle gaining mass to become a massive particle.

In our previous discussion of polarization optics, we have used the attenuation matrix of equation (10.6) which enables us to formulate the Lorentz group for the Stokes parameters [12]. Furthermore, this attenuation matrix makes it possible to make a continuous transformation from one matrix to another by adjusting the attenuation parameters in optical media. It could be interesting to design optical experiments along this direction.

In section 11.1 we discuss the issue of polarization optics and introduce the problem of decoherence between two beams. In section 11.2 we discuss the properties of the coherency matrix and the Stokes parameters as the elements in the coherency matrix. In section 11.3 we introduce the Poincaré sphere. In section 11.4 we formulate the entropy problem in terms of the coherency matrix. In section 11.5 we discuss further symmetries obtainable from the Poincaré sphere.

11.1 Polarization optics and decoherence

We saw in chapter 10 that the Jones vector formalism is a concrete physical example of the two-by-two representation of the Lorentz group. There, we were able to identify equation (10.6) as the attenuation matrix, equation (10.8) as the phase shifter matrix, and equation (10.10) as the rotation matrix, with appropriate transformation matrices of the Lorentz group. This formulation is not restricted solely to polarization optics, but can be applied to all two-beam systems with coherent or partially coherent phases, for example interferometers [13]. The Jones vectors cannot deal with the lack of coherence between two beams. However, the density matrix, the Stokes parameters, and the Poincaré sphere can deal with this problem [3].

For example, we can start with a pair of complex numbers a and b and construct the density matrix:

$$\rho = \begin{pmatrix} aa^* & ab^*e^{-\lambda t} \\ a^*be^{-\lambda t} & bb^* \end{pmatrix}. \tag{11.1}$$

The decay in the off-diagonal elements of this matrix play a fundamental role in the decoherence process. The determinant of this matrix is

$$aa^*bb^*(1 - e^{-2\lambda t}). \tag{11.2}$$

When $t = 0$, the system is in a pure state, and the determinant is zero, however as t increases, the determinant in equation (11.2) increases from zero to aa^*bb^*, and the system looses coherence.

Therefore, to address the issue of coherency between two beams, we need the two-by-two coherency matrix which takes the form of a density matrix. The elements of this matrix consist of four parameters which are then possible to be defined in terms of the Stokes parameters. It is the four Stokes parameters which define the parameters for the three-dimensional Poincaré sphere. However, the traditional Poincaré sphere needs only three parameters like the Euler angles. It is therefore necessary to investigate the role of the fourth Stokes parameter. We shall show here that the radius of the Poincaré sphere can change.

The Stokes parameters can be constructed from the two-component Jones vectors, which transform like the $SL(2, c)$ spinors. Thus the two-by-two coherency matrix should transform like the two-by-two form of the space–time four-vector discussed extensively in section 2.3. From this we know that the Lorentz group should leave the determinant of the coherency matrix invariant. However, the determinant of the coherency matrix changes according to the degree of coherency. Thus, the coherency is an extra-Lorentzian variable. We shall study its implications in Einstein's energy–momentum relation where the particle mass is a Lorentz-invariant quantity. Thus the symmetry group has to be extended to the $O(3, 2)$ group which was discussed in chapter 7.

11.2 Coherency matrix and Stokes parameters

We study first the question of coherency between the two orthogonal electric fields. Then, like the density matrix, the coherency matrix can be defined as [6, 7]

$$C = \begin{pmatrix} S_{11} & S_{12} \\ S_{21} & S_{22} \end{pmatrix}. \tag{11.3}$$

We can then write the elements of the matrix equation (11.3) as [11]

$$
\begin{aligned}
S_{11} &= \langle \psi_1^* \psi_1 \rangle = a^2, & S_{12} &= \langle \psi_1^* \psi_2 \rangle = ab\, e^{-(\lambda t + i\phi)}, \\
S_{21} &= \langle \psi_2^* \psi_1 \rangle = ab\, e^{-(\lambda t - i\phi)}, & S_{22} &= \langle \psi_2^* \psi_2 \rangle = b^2.
\end{aligned}
\tag{11.4}
$$

The absolute values of ψ_1 and ψ_2, respectively, are given by the diagonal elements. If the two transverse components are not completely coherent, the off-diagonal elements could be smaller than the product of ψ_1 and ψ_2. The degree of decoherence in the system is specified by λt. The decoherence is minimum if t is zero, and becomes maximum if t attains very large values.

Starting with the Jones vector of the form of equation (10.2), we can associate the elements of the coherency matrix as

$$
\begin{aligned}
S_{11} &= \langle E_x^* E_x \rangle, & S_{22} &= \langle E_y^* E_y \rangle, \\
S_{12} &= \langle E_x^* E_y \rangle, & S_{21} &= \langle E_y^* E_x \rangle.
\end{aligned}
\tag{11.5}
$$

Then the coherency matrix becomes

$$C = \begin{pmatrix} A^2 & AB\, e^{-(\lambda t + i\phi)} \\ AB\, e^{-(\lambda t - i\phi)} & B^2 \end{pmatrix}. \tag{11.6}$$

The symmetry properties of this matrix are what is of interest here. We found in chapter 10 that the transformation matrix applicable to the Jones vector is the two-by-two representation of the Lorentz group. Thus we are particularly interested in what transformation matrices are applicable to this coherency matrix.

The determinant of the above coherency matrix is calculated to be

$$\det(C) = (AB)^2 (1 - e^{-2\lambda t}), \tag{11.7}$$

and its trace is obtained as

$$\mathrm{tr}(C) = A^2 + B^2. \tag{11.8}$$

Now, we define the degree of polarization as [18]

$$f(t) = \sqrt{1 - \frac{4\det(C)}{(\operatorname{tr}(C))^2}} = \sqrt{1 - \frac{4(AB)^2(1 - e^{-2\lambda t})}{(A^2 + B^2)^2}}. \tag{11.9}$$

When $t = 0$, this degree becomes 1. When t attains very large values, the degree becomes

$$f(t)|_{t \to \infty} = \frac{A^2 - B^2}{A^2 + B^2}. \tag{11.10}$$

We can assume without loss of generality that A is greater than B. Should they be equal, the minimum degree of polarization is zero. If we then use the Lorentz transformation defined for the Jones vectors of section 10.1.3, this coherency matrix becomes transformed as

$$C' = G\,C\,G^\dagger = \begin{pmatrix} S'_{11} & S'_{12} \\ S'_{21} & S'_{22} \end{pmatrix}$$
$$= \begin{pmatrix} \alpha & \beta \\ \gamma & \delta \end{pmatrix}\begin{pmatrix} S_{11} & S_{12} \\ S_{21} & S_{22} \end{pmatrix}\begin{pmatrix} \alpha^* & \gamma^* \\ \beta^* & \delta^* \end{pmatrix}. \tag{11.11}$$

This is the same Lorentz transformation as was defined in section 2.2, which leads to the four-by-four transformation [3]

$$\begin{pmatrix} S'_{11} \\ S'_{22} \\ S'_{12} \\ S'_{21} \end{pmatrix} = \begin{pmatrix} \alpha^*\alpha & \gamma^*\beta & \gamma^*\alpha & \alpha^*\beta \\ \beta^*\gamma & \delta^*\delta & \delta^*\gamma & \beta^*\delta \\ \beta^*\alpha & \delta^*\alpha & \beta^*\beta & \delta^*\beta \\ \alpha^*\gamma & \gamma^*\gamma & \alpha^*\delta & \gamma^*\delta \end{pmatrix}\begin{pmatrix} S_{11} \\ S_{22} \\ S_{12} \\ S_{21} \end{pmatrix}. \tag{11.12}$$

From the components of the coherency matrix, it will prove to be useful to introduce four quantities as

$$S_0 = \frac{1}{2}(S_{11} + S_{22}), \qquad S_3 = \frac{1}{2}(S_{11} - S_{22}),$$
$$S_1 = \frac{1}{2}(S_{12} + S_{21}), \qquad S_2 = \frac{i}{2}(S_{12} - S_{21}). \tag{11.13}$$

They are called Stokes parameters, and the four-by-four transformation matrices applicable to these parameters are widely known as Mueller matrices [1, 7, 17]. These matrices are responsible for the relationship between polarization states of the incident light and the emerging light after passing through any number and any type of optical elements, such as polarizers, waveplates, and scatterers. It is by employing the Mueller matrices that we can perform Lorentz transformations on the four Stokes parameters. From there on we see that Stokes parameters indeed behave as the components of a four-vector.

Now, using equations (11.6) and (11.13), the expression of this four-vector is found to be

$$
\begin{pmatrix} S_0 \\ S_1 \\ S_2 \\ S_3 \end{pmatrix} = \frac{1}{2} \begin{pmatrix} A^2 + B^2 \\ 2AB(\cos\phi)e^{-\lambda t} \\ 2AB(\sin\phi)e^{-\lambda t} \\ A^2 - B^2 \end{pmatrix}.
\tag{11.14}
$$

Conversely, the coherency matrix can succinctly be constructed from Stokes parameters and the Pauli spin matrices as

$$
C = \sum_{i=0}^{3} \sigma_i \, S_i,
\tag{11.15}
$$

where σ_0 is the two-by-two identity matrix.

11.3 Poincaré sphere

Originally the Poincaré sphere proposed by Henri Poincaré was a three-dimensional object for representing the polarization state of light. The three Stokes parameters S_1, S_2, S_3 correspond to the rectangular Cartesian coordinates. The radius of the Poincaré sphere is given in terms of these three Stokes parameters as

$$
r = \sqrt{S_1^2 + S_2^2 + S_3^2}.
\tag{11.16}
$$

While S_0 is known to represent the intensity of the light, it is now well established that it also enjoys being the time-like component of a four-vector. We shall use this property of the Stokes vector to construct a second concentric sphere, whose radius is given by $s = S_0$. Within the framework of the Lorentz group the relation between these two radii, namely $s^2 - r^2$, is an invariant. This implies that the determinant of the density matrix is also subject to the same restriction, which cannot accommodate the decoherence process through the decaying of the off-diagonal elements. Therefore one is tempted to ask whether it is possible to formulate this process by introducing the symmetry properties of a larger Lorentz group, namely $O(3, 2)$, the de Sitter group. In the following subsections we shall discuss and answer these questions.

11.3.1 Two concentric Poincaré spheres

We can combine equations (11.14) and (11.16) to rewrite the radius of the sphere as

$$
r = \frac{1}{2}\sqrt{(A^2 - B^2)^2 + 4(AB)^2 e^{-2\lambda t}},
\tag{11.17}
$$

which describes the conventional Poincaré sphere. However, the four-vector in equation (11.14), apart from the space-like components, has a time-like component which provides another radius

$$s = \frac{1}{2}(A^2 + B^2) \tag{11.18}$$

defining the concentric outer sphere. The quantity $s^2 - r^2$ is Lorentz-invariant and is equal to the value of the determinant in equation (11.7). The t parameter cannot be changed in the Lorentzian regime. However, this restriction can be released, if a larger Lorentz group is introduced, as we shall see in the next section. In that case, when $t = 0$, the inner radius is equal to the outer radius, and it becomes

$$S_3 = \frac{1}{2}(A^2 - B^2), \tag{11.19}$$

when t attains very large values.

Let us introduce a spherical coordinate system with

$$r_z = (A^2 - B^2)/2 = r(\cos \theta),$$
$$r_x = AB(\cos \phi)e^{-\lambda t} = r(\sin \theta)\cos \phi, \tag{11.20}$$
$$r_y = AB(\sin \phi)e^{-\lambda t} = r(\sin \theta)\sin \phi.$$

Since, the Lorentz symmetry allows rotations in this three-dimensional scheme, with an appropriate rotation the four-vector can be brought to

$$\begin{pmatrix} s \\ r \\ 0 \\ 0 \end{pmatrix}. \tag{11.21}$$

The rotations do not change the radii of the outer and inner spheres, thus s and r remain invariant. On the other hand, if the above four-vector is boosted with the inverse of the matrix in equation (2.10), it becomes

$$\begin{pmatrix} s(\cosh \eta) - r(\sinh \eta) \\ r(\cosh \eta) - s(\sinh \eta) \\ 0 \\ 0 \end{pmatrix}. \tag{11.22}$$

This transformation changes the outer and inner radii, but keeps $(s^2 - r^2)$ invariant, as can be seen from

$$[s(\cosh \eta) - r(\sinh \eta)]^2 - [r(\cosh \eta) - s(\sinh \eta)]^2 = s^2 - r^2. \tag{11.23}$$

One can choose the value of η such that

$$r(\cosh \eta) - s(\sinh \eta) = 0, \tag{11.24}$$

which leads to $\tanh \eta = r/s$. If this condition is met, the four-vector of equation (11.22) becomes

$$\begin{pmatrix} \sqrt{s^2 - r^2} \\ 0 \\ 0 \\ 0 \end{pmatrix} = \begin{pmatrix} AB\sqrt{1 - e^{-2\lambda t}} \\ 0 \\ 0 \\ 0 \end{pmatrix}. \tag{11.25}$$

The Lorentz symmetry indeed brings the Poincaré sphere to a one-number system. How can it be possible to change the value of $(s^2 - r^2)$ in the above expression by changing the time variable t, while it is clearly an invariant quantity in the $O(3, 1)$ regime? In the next section we shall see that this can be achieved when the symmetry group is enlarged.

11.3.2 $O(3, 2)$ symmetry of the Poincaré sphere

In order to deal with the problem of invariance of coherence in the $O(3, 1)$ regime, we introduce a larger group, namely the $O(3, 2)$ de Sitter space with (t, u, z, x, y) variables while allowing rotations

$$\begin{pmatrix} \cos\chi & -\sin\chi & 0 & 0 & 0 \\ \sin\chi & \cos\chi & 0 & 0 & 0 \\ 0 & 0 & 1 & 0 & 0 \\ 0 & 0 & 0 & 1 & 0 \\ 0 & 0 & 0 & 0 & 1 \end{pmatrix}, \tag{11.26}$$

between the two time-like components t and u. The invariant quantity in this space is

$$t^2 + u^2 - z^2 - x^2 - y^2. \tag{11.27}$$

The de Sitter space contains two Minkowskian subspaces, namely the space of (t, z, x, y) with the invariant of $(t^2 - z^2 - x^2 - y^2)$, and similarly, of (u, z, x, y) with the invariant of $(u^2 - z^2 - x^2 - y^2)$.

As can be seen from equation (11.27), if $z = x = y = 0$, this quantity is $(t^2 + u^2) = k^2$ and remains as an invariant in this space.

Let us consider the five-vector $(k, 0, 0, 0, 0)$ in the de Sitter space. The above five-by-five matrix changes this five-vector to

$$(k(\cos\chi), k(\sin\chi), 0, 0, 0). \tag{11.28}$$

Thus, in the Minkowskian world of (t, z, x, y), the invariant quantity is $k^2 \cos^2\chi$, and $k^2 \sin^2\chi$ in the Minkowskian space of (u, z, x, y), where now the four-vectors in these spaces are

$$(k(\cos\chi), 0, 0, 0) \quad \text{and} \quad (k(\sin\chi), 0, 0, 0), \tag{11.29}$$

respectively.

Let us compare the second four-vector of equation (11.29) with the four-vector of equation (11.25). If we identify the parameter $k(\sin\chi)$ in equation (11.29) with $\sqrt{s^2 - r^2}$ of equation (11.25), we have

$$s^2 - r^2 = k^2 \sin^2 \chi. \tag{11.30}$$

This further allows us to identify k as AB in equation (11.25) and

$$(AB)^2 (\sin \chi)^2 = (AB)^2 (1 - e^{-2\lambda t}), \tag{11.31}$$

which leads to

$$\cos \chi = e^{-\lambda t}. \tag{11.32}$$

Here, we note that $e^{-\lambda t}$ should always be smaller than 1. Thus, we can conclude that the decoherence parameter can be identified with the angle variable χ in the de Sitter space. This angle will be called the *decoherency angle*. When $\chi = 0°$, the decoherence is minimum, and becomes maximum when $\chi = 90°$.

After changing the t variable, we can make inverse transformations to return to the four-vector of the form given in equation (11.14). Indeed, it is gratifying to note that we now have the freedom of changing this time variable with a symmetry operation. In terms of this symmetry parameter, we can write the coherency matrix of equation (11.6) as

$$C = \begin{pmatrix} A^2 & AB\,e^{-i\phi}(\cos \chi) \\ AB\,e^{i\phi}(\cos \chi) & B^2 \end{pmatrix}. \tag{11.33}$$

The determinant of this matrix is $(AB)^2 \sin^2 \chi$. If $\chi = 0°$ and $t = 0$, the system is in a pure state. As t becomes large, the angle χ approaches 90°. Therefore, the de Sitter parameter χ neatly takes care of the loss of coherence in the two-beam system. This aspect of coherency is tightly connected to the variable radius of the Poincaré sphere as illustrated in figure 11.1.

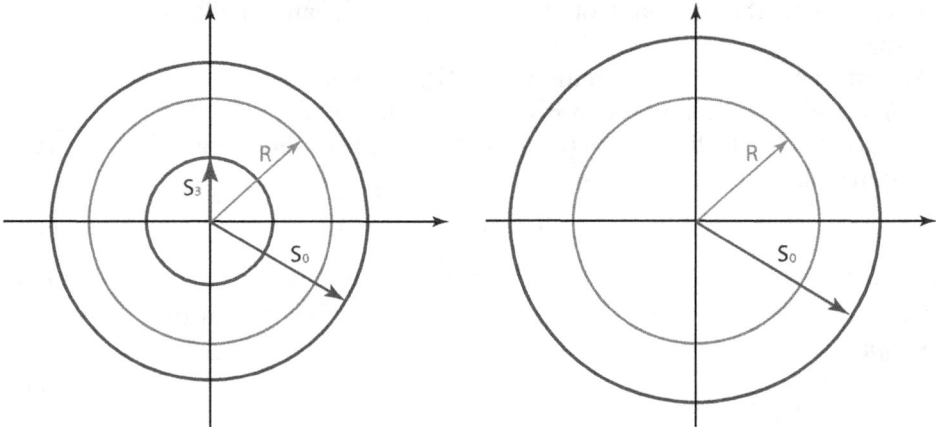

Figure 11.1. Variable radius of the Poincaré sphere. From equations (11.17) and (11.32), the variable radius R takes its maximum value S_0, when $\chi = 0°$. It becomes minimum when the coherency angle reaches 90°. Its minimum value is S_3 as is illustrated in the left figure. The degree of polarization is maximum when $R = S_0$ and is minimum when $R = S_3$. According to equation (11.19), S_3 becomes 0 when $A = B$, thus the minimum value of R becomes zero, as is indicated in the right figure. Its maximum value is still S_0 [14].

Since we can have two separate Minkowskian spaces, our analysis for the second space should exactly be the same, except that now $\cos \chi$ is replaced with $\sin \chi$. Now, the coherency matrix is written as

$$C' = \begin{pmatrix} A^2 & AB \, e^{-i\phi}(\sin \chi) \\ AB \, e^{i\phi}(\sin \chi) & B^2 \end{pmatrix}. \tag{11.34}$$

The determinant of this matrix is $(AB)^2 \cos^2 \chi$. The sum of the determinants of C and C' is

$$\det(C) + \det(C') = (AB)^2 \tag{11.35}$$

and is independent of the parameter χ. We can observe that, these two Lorentzian spaces are coupled in a Pythagorean mode.

11.3.3 The Poincaré circle

As we saw in chapter 10, it is possible to construct the Jones vector of equation (10.2) by making Lorentz transformations to obtain a simpler form

$$\begin{pmatrix} E_x \\ E_y \end{pmatrix} = \begin{pmatrix} A\exp\{i(kz - \omega t + \phi_1)\} \\ A\exp\{i(kz - \omega t + \phi_2)\} \end{pmatrix}. \tag{11.36}$$

Hence, we can now drop the amplitude A in equation (11.33) and work with the coherency matrix of the form

$$C = \begin{pmatrix} 1 & e^{-i\phi} \cos \chi \\ e^{i\phi} \cos \chi & 1 \end{pmatrix}. \tag{11.37}$$

By using the correspondence between the matrix elements in the coherence matrix given in equation (11.13), the Stokes parameters can be written as

$$\begin{aligned} S_0 &= 1, \qquad S_3 = 0, \\ S_1 &= (\cos \chi)\cos \phi, \qquad S_2 = (\cos \chi)\sin \phi. \end{aligned} \tag{11.38}$$

Now, since $S_3 = 0$, the Poincaré sphere dimensionally reduces and becomes a circle with radius

$$r_2 = \sqrt{S_1^2 + S_2^2} = \cos \chi. \tag{11.39}$$

The determinant $\sin^2 \chi$ of C in equation (11.37) would remain invariant in the Lorentzian regime, as is the invariance of the Stokes parameters of equation (11.38)

$$S_0^2 - S_1^2 - S_2^2 = \sin^2 \chi. \tag{11.40}$$

The coherency parameter depends on the environment and thus, in that sense, is not a fundamental quantity. However, when equation (2.42) is taken into account, in this particular case its mathematical role can be considered to be similar to the mass of the particle. We illustrate this aspect of the decoherency in table 11.1.

Table 11.1. Polarization optics and special relativity sharing similar mathematics. Each matrix has its clear role in both optics and relativity. The determinant of the two-by-two matrix obtained from the Stokes vector or from the four-momentum remains invariant under Lorentz transformations. It is interesting to note that the decoherency parameter (least fundamental) in optics corresponds to the mass (most fundamental) in particle physics.

Polarization optics	Transformation matrix	Particle symmetry
Phase shift ϕ	$\begin{pmatrix} e^{-i\phi/2} & 0 \\ 0 & e^{i\phi/2} \end{pmatrix}$	Rotation around z
Rotation around z	$\begin{pmatrix} \cos(\theta/2) & -\sin(\theta/2) \\ \sin(\theta/2) & \cos(\theta/2) \end{pmatrix}$	Rotation around y
Squeeze along x and y	$\begin{pmatrix} e^{\mu/2} & 0 \\ 0 & e^{-\mu/2} \end{pmatrix}$	Boost along z
$\sin^2\chi$	Determinant	$(\text{mass})^2$

11.3.4 Diagonalization of the coherency matrix

If the coherency matrix is divided by 2, it can also serve as the density matrix. We can therefore write

$$\rho(\chi) = \frac{C}{2} = \frac{1}{2}\begin{pmatrix} 1 & e^{-i\phi}\cos\chi \\ e^{i\phi}\cos\chi & 1 \end{pmatrix}. \tag{11.41}$$

This density matrix has a trace of 1.

We can then diagonalize this density matrix to

$$\rho(\chi) = \frac{1}{2}\begin{pmatrix} 1 + \cos\chi & 0 \\ 0 & 1 - \cos\chi \end{pmatrix}. \tag{11.42}$$

Let us exploit the symmetry between $\cos\chi$ and $\sin\chi$, and write another coherency matrix for the other space, where we have

$$\rho'(\chi) = \frac{C'}{2} = \frac{1}{2}\begin{pmatrix} 1 & e^{-i\phi}\sin\chi \\ e^{i\phi}\sin\chi & 1 \end{pmatrix}, \tag{11.43}$$

whose diagonalized form becomes

$$\rho'(\chi) = \frac{1}{2}\begin{pmatrix} 1 + \sin\chi & 0 \\ 0 & 1 - \sin\chi \end{pmatrix}. \tag{11.44}$$

These two matrices are mathematically equivalent. The choice of which to use can be determined by their entropy particularities as will be discussed in the next section.

11.4 The entropy problem

The entropy

$$S = -\text{Tr}(\rho \ln \rho) \tag{11.45}$$

for the density matrix of equation (11.42) can be written as [9, 15, 16]

$$S(\chi) = -\frac{1 + \cos \chi}{2} \ln \left(\frac{1 + \cos \chi}{2} \right) - \frac{1 - \cos \chi}{2} \ln \left(\frac{1 - \cos \chi}{2} \right). \tag{11.46}$$

This can be simplified to

$$S(\chi) = -[\cos^2(\chi/2)]\ln[\cos^2(\chi/2)] - [\sin^2(\chi/2)]\ln[\sin^2(\chi/2)]. \tag{11.47}$$

From this we see that if the entropy is zero the system is completely coherent with $\chi = 0°$. When the system is totally incoherent with $\chi = 90°$ the entropy takes the maximum value of $\ln(2)$.

In a similar manner, the entropy of the second space can be expressed as

$$S'(\chi) = -\frac{1 + \sin \chi}{2} \ln \left(\frac{1 + \sin \chi}{2} \right) - \frac{1 - \sin \chi}{2} \ln \left(\frac{1 - \sin \chi}{2} \right). \tag{11.48}$$

The entropy S of the first space is a monotonically increasing function of χ, while that of the second space S' is a decreasing function. In chapter 8, we have discussed Feynman's rest of the Universe in detail within the context of coupled harmonic oscillators. The first of those belongs to the part of the Universe in which we make measurements, while the second is the rest of the Universe. The decrease of entropy in the second space can be interpreted as the gain of energy in that space. There is a decoherence and a recoherence process going on in these coupled universes. Yet, the sum of the entropies is not a conserved quantity of the total system. Nevertheless, if we insist on the conservation of some physical quantity, we can resort to equation (11.35), namely the sum of the determinants of the two coherency or density matrices, which turns out to be independent of the coherency angle χ.

11.5 Further symmetries from the Poincaré sphere

In section 11.4 it was shown that the Poincaré sphere contains the symmetry of the Lorentz group applicable to the momentum–energy four-vector. It is true that the Lorentz group cannot tolerate a variable mass, yet the Poincaré sphere has an extra variable χ which we can exploit to be able to manage the mass value of this four-vector.

11.5.1 Momentum four-vector and the Poincaré sphere

Let us now write the four-momentum matrix as

$$P = \begin{pmatrix} E + p & 0 \\ 0 & E - p \end{pmatrix}. \tag{11.49}$$

Here we let the particle move along the z-direction, thus E and p are the energy and the magnitude of the momentum, respectively.

If we then write $p = E \cos \chi$, the matrix P becomes

$$P = E \begin{pmatrix} 1 + \cos \chi & 0 \\ 0 & 1 - \cos \chi \end{pmatrix}, \qquad (11.50)$$

which is like the matrix C of equation (11.42). If instead, $p = E \sin \chi$, then

$$P' = E \begin{pmatrix} 1 + \sin \chi & 0 \\ 0 & 1 - \sin \chi \end{pmatrix}, \qquad (11.51)$$

which is like the C' matrix of equation (11.44).

If we then choose P' of equation (11.51), we see that if $\chi = 0°$, the matrix becomes the four-momentum of the particle at rest with a mass E. On the other hand, if $\chi = 90°$, the particle becomes massless with a momentum of E.

If we start from a massive particle at rest, we would like to reach a massless particle with the same energy. This problem was discussed in detail in chapter 3 within the framework of the Lorentz group. The transition from the massive particle to the massless particle was shown to be a singular transformation, since the mass remains invariant under Lorentz transformations. This transition can only be accomplished with a tangential continuity as discussed in sections 3.4 and 3.5.

If instead we consider the Poincaré sphere, this transition is continuous as indicated in figure 11.2. The question is whether this extra-Lorentzian transformation can be accommodated by a larger symmetry group.

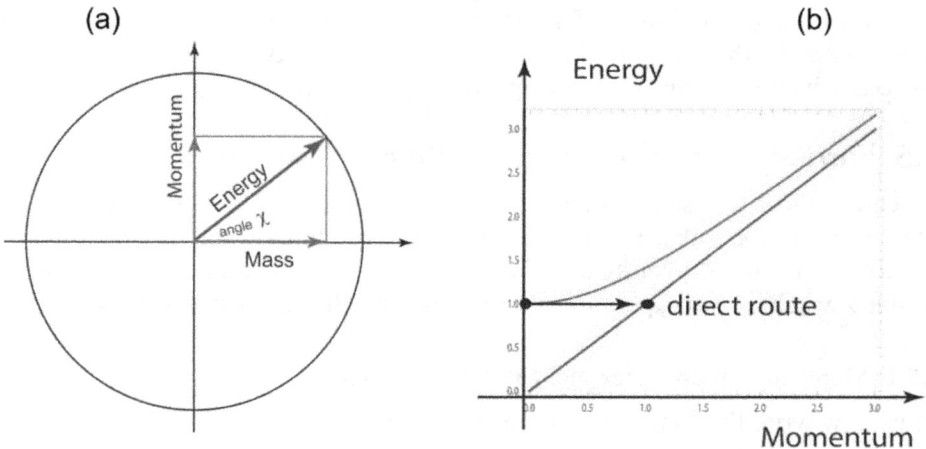

Figure 11.2. The momentum–mass relation for a fixed value of energy in (a). As the angle χ increases, the momentum increases, while the mass becomes smaller. The particle becomes massless for $\chi = 90°$. This transition is described in terms of the traditional energy–momentum plot in (b). This transition is not allowed within the framework of the Lorentz group. The Poincaré sphere allows this extra-Lorentz transformation [5].

11.5.2 Mass variation within $O(3, 2)$ symmetry

We have already seen that although the original Poincaré sphere has one fixed radius, this radius can change depending on the degree of coherence. In this section this change of radius will be associated to the change of mass of the particle.

In chapter 7, it was shown that the group $O(3, 2)$ is the Lorentz group applicable to a five-dimensional space which contains three space dimensions and two time dimensions. Therefore, this leads to two energy variables, and the five-component vector becomes

$$(E_1, E_2, p_z, p_x, p_y). \tag{11.52}$$

In studying this group, we would have to use five-by-five matrices. Since we are interested in the subgroups of $O(3, 2)$ this is not necessary. We note first that there is a three-dimensional Euclidean space consisting of p_z, p_x, and p_y. The $O(3)$ rotation group is applicable to this space. This is the same as was true for the $O(3, 1)$ Lorentz group, applicable to three space dimensions and one time dimension.

Since the momentum is in the z-direction, the five-vector given in equation (11.52) becomes

$$(E_1, E_2, p, 0, 0). \tag{11.53}$$

These two energy variables then take the form

$$E_1 = \sqrt{p^2 + m_1^2} \quad \text{and} \quad E_2 = \sqrt{p^2 + m_2^2}. \tag{11.54}$$

Here [3, 21] we have

$$m_1 = m \cos \chi \quad \text{and} \quad m_2 = m \sin \chi, \tag{11.55}$$

which results in

$$E_1^2 + E_2^2 = m^2 + 2p^2. \tag{11.56}$$

This value remains constant for a fixed value of p^2. This rotational symmetry in the two-dimensional space of E_1 and E_2 defines the $O(3, 2)$ de Sitter symmetry [3, 4].

The two Lorentz subgroups applicable to the Minkowskian spaces of

$$(E_1, p, 0, 0) \quad \text{and} \quad (E_2, p, 0, 0), \tag{11.57}$$

are the most important subgroups for our purpose here. In the two-by-two matrix representation, these four-momenta can be written as

$$\begin{pmatrix} E_1 + p & 0 \\ 0 & E_1 - p \end{pmatrix} \quad \text{and} \quad \begin{pmatrix} E_2 + p & 0 \\ 0 & E_2 - p \end{pmatrix}. \tag{11.58}$$

The determinants of these matrices are equal to $m^2 \cos^2 \chi$ and $m^2 \sin^2 \chi$, respectively.

We concentrate now only on the matrix with E_1. It takes a maximum value of $\sqrt{p^2 + m^2}$ when $\chi = 0°$ and $E_1 = p$. The de Sitter symmetry allows us to jump from one mass hyperbola to another in this fixed-momentum variation. This is illustrated in figure 11.3.

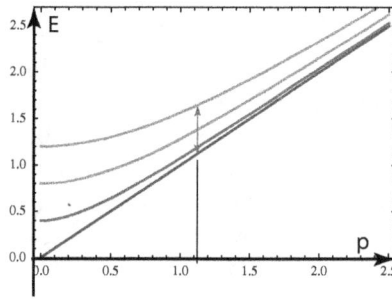

Figure 11.3. Energy–momentum hyperbolas for different values of the mass. The Lorentz group does not allow us to jump from one hyperbola to another, but it is possible within the framework of the $O(3, 2)$ de Sitter symmetry. This figure illustrates the transition while the magnitude of the momentum is kept constant [5].

The symmetry group $O(3, 2)$ or $SO(3, 2)$ appears in various branches of physics, ranging from quantum mechanics to extended theories of gravity [20, 21]. The quantum mechanical case in the context of the Dirac's harmonic oscillator [8] was extensively discussed in section 7.6. Additionally, spin–orbit coupled harmonic oscillators also admit this same symmetry group [10].

Emerging from different viewpoints, attempts to accommodate another time-like dimension to the usual space–time is, in fact, not new [2, 19]. The isometry group for the AdS^4 (anti-de Sitter) space–time is accommodated by $SO(3, 2)$. This group allows closed time-like curves. It solves Einstein's equations with a negative cosmological constant accounting for a contracting universe.

Similarly, the dS^4 (de Sitter) space–time whose isometry group is $O(4, 1)$, in contrast with AdS^4, solves Einstein's equations with a positive cosmological constant. This accounts for the expansion of the Universe which is what we are observing now in the real cosmos [21]. It is gratifying to note that the Poincaré sphere contains this important symmetry.

References

[1] Azzam R M A-G and Bashara N M 1999 *Ellipsometry and Polarized Light* (Amsterdam: Elsevier)

[2] Bars I, Deliduman C, and Andreev O 1998 Gauged duality, conformal symmetry, and spacetime with two times *Phys. Rev.* D **58** 066004

[3] Başkal S and Kim Y S 2006 de Sitter group as a symmetry for optical decoherence *J. Phys. A: Math. Gen.* **39** 7775–88

[4] Başkal S and Kim Y S 2013 Lorentz group in ray and polarization optics *Mathematical Optics: Classical, Quantum and Computational Methods* ed V Lakshminarayanan, M L Calvo and T Alieva (Boca Raton, FL: Taylor and Francis) pp 303–49

[5] Başkal S, Kim Y S, and Noz M E 2015 *Physics of the Lorentz Group* (Bristol: IOP Publishing)

[6] Born M and Wolf E 1999 *Principles of Optics: Electromagnetic Theory of Propagation, Interference and Diffraction of Light* 7th edn (Cambridge: Cambridge University Press)

[7] Brosseau C 1998 *Fundamentals of Polarized Light: A Statistical Optics Approach* (New York: Wiley)

[8] Dirac P A M 1963 A remarkable representation of the 3 + 2 de Sitter group *J. Math. Phys.* **4** 901–9

[9] Eisert J, Cramer M, and Plenio M B 2010 *Colloquium*: area laws for the entanglement entropy *Rev. Mod. Phys.* **82** 277–306

[10] Haaker S M, Bais F A, and Schoutens K 2014 Noncompact dynamical symmetry of a spin-orbit-coupled oscillator *Phys. Rev.* A **89**

[11] Han D, Kim Y S, and Noz M E 1997 Jones-matrix formalism as a representation of the Lorentz group *J. Opt. Soc. Am.* A **14** 2290–8

[12] Han D, Kim Y S, and Noz M E 1997 Stokes parameters as a Minkowskian four-vector *Phys. Rev.* E **56** 6065–76

[13] Han D, Kim Y S, and Noz M E 2000 Interferometers and decoherence matrices *Phys. Rev.* E **61** 5907–13

[14] Kim Y S and Noz M E 2013 Symmetries shared by the Poincaré group and the Poincaré sphere *Symmetry* **5** 233–52

[15] Kim Y S and Noz M E 2014 Entropy and temperature from entangled space and time *Phys. Sci. Int. J.* **4** 1015–39

[16] Kim Y S and Wigner E P 1990 Entropy and Lorentz transformations *Phys. Lett.* A **147** 343–7

[17] Mueller H 1943 Memorandum on the polarization optics of the photo elastic shutter *OSRD project OEMsr-576 2* Office of Scientific Research and Development of the United States

[18] Saleh B E A and Carl Teich M 2007 *Fundamentals of Photonics* (*Wiley Series in Pure and Applied Optics*) 2nd edn (Hoboken, NJ: Wiley)

[19] Wesson P S 2002 Five-dimensional relativity and two times *Phys. Lett.* B **538** 159–63

[20] Wesson P S 2006 *Five-dimensional Physics: Classical and Quantum Consequences of Kaluza–Klein Cosmology* (Singapore: World Scientific)

[21] Zee A 2013 *Einstein Gravity in a Nutshell. In a Nutshell* (Princeton, NJ: Princeton University Press)

IOP Publishing

Mathematical Devices for Optical Sciences

Sibel Başkal, Young S Kim, and Marilyn E Noz

Appendix A

Covariant harmonic oscillators and the quark–parton puzzle

One method of combining quantum mechanics and special relativity takes the form of efforts to construct bound-state relativistic wave functions with an appropriate probability interpretation. Because of its mathematical simplicity, the harmonic oscillator has played a key role in the developing stages of non-relativistic quantum mechanics, statistical mechanics, the theory of specific heat, molecular theory, quantum field theory, the theory of superconductivity, the theory of coherent light, and many others. Thus it is quite natural that the first covariant wave function be a covariant harmonic oscillator wave function.

Another advantage of the harmonic oscillator formalism is that it is the basic language for phase space. Indeed, it is possible to construct a phase-space picture of the covariant harmonic oscillator formalism, as was discussed in sections 5.4 and 5.5. In this picture, we can see clearly how the time–energy uncertainty relation can be incorporated into the position–momentum relation in a covariant manner. The Lorentz transformation in this case is a canonical transformation in phase space.

As the size of the proton is 10^{-5} that of the hydrogen atom, it was not unnatural to assume that the proton is like a point charge in atomic physics. When, however, Hofstadter and McAllister [30] were performing experiments on electron scattering from proton targets, they observed that the proton charge is spread out. The effects of these deformations have been measured, and many experimental plots are available in terms of the proton speed [29].

In 1964, it was observed that the proton is the bound state of the more fundamental constituents called *quarks* [23]. Although the proton is different from the hydrogen atom, it inherits the same quantum mechanics from the hydrogen atom. Unlike the hydrogen atom, the proton can be accelerated, and its speed can become very close to that of light. Thus, it is possible to study the quantum mechanics of the hydrogen atom or bound states in the Lorentz-covariant world by studying the proton in the quark model.

To describe what happens to the probability distribution when the proton speed becomes very close to that of light, Feynman observed that the proton appears like a collection of *partons* with a wide-spread momentum distribution [17, 18]. Partons are, therefore, like free particles. Quarks and partons are the same particles but they appear differently to observers in two different reference frames. Therefore, there must be a Lorentz-covariant model for quantum bound states.

In section A.1 we discuss the harmonic oscillator formalism. In particular we discuss the many aspects of this formalism which can be used in many branches of physics including the quark–parton puzzle. In section A.1.1, we formulate the problem by writing down the relativistically invariant differential equation which leads to the covariant harmonic oscillator formalism.

In section A.1.2 we study solutions of the oscillator differential equation which are normalizable in the four-dimensional Minkowskian (t, z, x, y) space. We show here how the representations of the Poincaré group for massive hadrons are constructed from the normalizable harmonic oscillator wave functions. It is shown that they form the basis for unitary irreducible representations of the Poincaré group, as well as that for the $O(3)$-like little group for massive particles.

In subsection A.1.3 we study the Lorentz transformation properties of the harmonic oscillator wave functions. The linear unitary representation of Lorentz transformation is provided for the harmonic oscillator wave functions. In section A.1.4 we develop the phase-space picture for relativistic oscillators.

In section A.2 we discuss the quark–parton puzzle along with solutions to the problem using the Lorentz-covariant harmonic oscillator. In section A.2.1 we formulate the Lorentz-covariant quark model. In section A.2.2 we formulate Feynman's parton picture, show the contradictions it contains, and suggest how these contradictions can be resolved. In section A.2.3 the proton structure function is introduced and the proton form factor is derived and shown be a consequence of Lorentz coherence. In section A.2.4 we discuss Lorentz coherence in momentum–energy space. In section A.2.5 we show how the temperature of a fast moving hadron increases. This leads to the quarks being heated and appearing as partons.

A.1 The covariant harmonic oscillator

As early as 1945, Dirac suggested the use of normalizable relativistic oscillator wave functions to construct representations of the Lorentz group [11]. Dirac's plan to construct a relativistic dynamics of the atom using Poisson brackets is contained in his 1949 paper [12]. Here Dirac emphasizes that the task of constructing a relativistic dynamics is equivalent to constructing a representation of the Poincaré group. His approach to relativistic quantum mechanics allows us to construct relativistic bound-state wave functions which can be Lorentz-transformed.

In connection with relativistic particles with internal space–time structure, Yukawa attempted to construct relativistic harmonic oscillator wave functions in

1953 [69]. Yukawa observed that an attempt to solve a relativistic oscillator wave equation in general leads to infinite-component wave functions, and that finite-component wave functions may be chosen if a subsidiary condition involving the four-momentum of the particle is considered. This proposal of Yukawa was further developed by many authors [35, 55, 61–63].

The basic problem facing any relativistic harmonic oscillator equation is the negative-energy spectrum due to time-like excitations. However, it was thought that eliminating them would lead to a violation of probability conservation. It is now possible to construct harmonic oscillator wave functions without time-like wave functions. These wave functions form the vector spaces for unitary irreducible representations of the Poincaré group.

A.1.1 Differential equations of the covariant harmonic oscillator

In order explain new physical phenomena it is often necessary that a new set of mathematical formulae be developed. Dirac's approach [10] to solving this problem is that it is more profitable to construct plausible mathematical devices which can describe quantitatively the real world, and then add physical interpretations to the mathematical formalism. Both special relativity and quantum mechanics were developed in Dirac's way, and most of the new physical models these days are developed in this way.

In order to discuss covariant harmonic oscillators we consider a relativistic hadron consisting of two quarks bound together by a harmonic oscillator potential of unit strength. We start with the Lorentz-invariant differential equation of Feynman *et al* [20]:

$$\left\{ -2\left[\left(\frac{\partial}{\partial x_a^\mu} \right)^2 + \left(\frac{\partial}{\partial x_b^\mu} \right)^2 \right] + \left(\frac{1}{16} \right)(x_a^\mu - x_b^\mu)^2 + m_0^2 \right\} \phi(x_a^\mu, x_b^\mu) = 0, \qquad (A.1)$$

where x_a^μ and x_b^μ are space–time coordinates for the first and second quarks, respectively. It is convenient to introduce the following variables:

$$X = (x_a^\mu + x_b^\mu)/2 \quad \text{and} \quad x = (x_a^\mu - x_b^\mu)/2\sqrt{2}. \qquad (A.2)$$

The four-vector X specifies where the hadron is located in space–time, while the variable x measures the space–time separation between the quarks. Then equation (A.1) can be written as

$$\left(\frac{\partial^2}{\partial X_\mu^2} - m_0^2 + \frac{1}{2} \left[\frac{\partial^2}{\partial x_\mu^2} - x_\mu^2 \right] \right) \phi(X, x) = 0. \qquad (A.3)$$

These variables allow $\phi(X, x)$ to be separated into

$$\phi(X, x) = f(X)\psi(x). \qquad (A.4)$$

Here $f(X)$ satisfies the differential equation

$$\left\{ \frac{\partial^2}{\partial X_\mu^2} - m_0^2 - (\lambda + 1) \right\} f(X) = 0, \tag{A.5}$$

and $\psi(x)$ satisfies the following differential equation

$$\left\{ \frac{1}{2} \left[\frac{\partial^2}{\partial x_\mu^2} - x_\mu^2 \right] + (\lambda + 1) \right\} \psi(x) = 0. \tag{A.6}$$

Since equation (A.5) is a Klein–Gordon equation, its solution takes the form

$$f(X) = e^{\pm i P_\mu X^\mu}. \tag{A.7}$$

Here

$$-P^2 = -P_\mu P^\mu = M^2 = m_0^2 + (\lambda + 1),$$

where P and M are the four-momentum and mass of the hadron, respectively. The solution of equation (A.6) determines the eigenvalue λ.

The four-momenta of the quarks p_a and p_b can also be combined into the total four-momentum and momentum–energy separation between the quarks:

$$P = p_a + p_b \quad \text{and} \quad q = \sqrt{2}(p_a - p_b). \tag{A.8}$$

Again P is the hadronic four-momentum conjugate to X. If there exist wave functions which can be Fourier-transformed then the internal momentum–energy separation q is conjugate to x. The differential equation in the q space is the same as the harmonic oscillator equation for the x-space given in equation (A.6) provided that the momentum–energy wave functions can be obtained from the Fourier transformation of the space–time wave function.

A.1.2 Normalizable solutions of the relativistic oscillator equations

Since there are many different forms of solutions of the Lorentz-invariant oscillator equation of equation (A.6), there are many papers in the literature reflecting the view that the solution of this equation has to be Lorentz- invariant [20]. The requirement that $(t^2 - z^2 - x^2 - y^2)$ be invariant leads to wave functions which are not normalizable. We shall discuss later Dirac's [11] normalizable solutions which are covariant but not invariant. These wave functions appear different to observers in different Lorentz frames [40, 43].

When the hadron moves along the Z-direction then the hadronic factor $f(X)$ of equation (A.7) is Lorentz-transformed in the same manner as scalar fields are transformed. Since the Z-direction is also the z-direction, the Lorentz transformation of the internal coordinates from the laboratory frame to the hadronic rest frame takes the form

$$t' = (t - \beta z)/(1 - \beta^2)^{1/2},$$
$$z' = (z - \beta t)/(1 - \beta^2)^{1/2}, \qquad \text{(A.9)}$$
$$x' = x, \qquad y' = y,$$

where β is the velocity of the hadron moving along the z-direction and the primed quantities are the coordinate variables in the hadronic rest frame.

Using these primed variables, the oscillator differential equation for the moving frame becomes

$$\frac{1}{2}\left\{-\nabla'^2 + \frac{\partial^2}{\partial t'^2}\left((t')^2 - (\vec{x}')^2\right)\right\}\psi(x) = (\lambda + 1)\psi(x). \qquad \text{(A.10)}$$

The solution of this equation is

$$\psi_\eta(x) = \left(\frac{1}{\pi}\right)^2 \left(\frac{1}{2}\right)^{(k+n+a+b)/2} \left(\frac{1}{k!n!a!b!}\right)^{1/2} H_k(t')H_n(z')$$
$$\times H_a(x')H_b(y')e^{-\frac{1}{2}\left(t'^2+z'^2+x'^2+y'^2\right)}, \qquad \text{(A.11)}$$

where k, n, a, and b are integers, and $H_k(t')$, $H_n(z')$, $\cdots$ are the Hermite polynomials. Thus this is a product of four one-dimensional oscillator wave functions. This wave function is normalizable, but the eigenvalue is

$$\lambda = (k - n - a - b). \qquad \text{(A.12)}$$

There are therefore, infinitely many possible combinations of k, n, a, and b, for a given finite value of λ, and thus the most general solution of the oscillator differential is infinitely degenerate [69].

Giving physical interpretations to infinite-component wave functions is very difficult, if not impossible. Therefore we seek a finite set from the infinite number of wave functions at least in one Lorentz frame. To do this we invoke the restriction that there are no time-like oscillations in the Lorentz frame in which the hadron is at rest. To do this we set $k = 0$ in equations (A.11) and (A.12). We are then led to ask the questions:
1. Is it possible to give physical interpretations to the wave functions belonging to the resulting finite set?
2. Is it still possible to maintain Lorentz covariance with this condition?

Physical interpretation for the first question will be given in section A.1.2. Therefore let us examine the second question here.

For a hadron traveling along the z-axis, the $k = 0$ condition is equivalent to

$$\left(\frac{\partial}{\partial t'} + t'\right)\psi_\eta(x') = 0. \qquad \text{(A.13)}$$

The above condition results in the general form

$$P_\mu\left(x^\mu - \frac{\partial}{\partial x_\mu}\right)\psi_\eta(x) = 0. \qquad \text{(A.14)}$$

This means that the $k = 0$ condition is covariant [32, 40]. The wave function belonging to this finite set can be written as

$$\psi_\eta^{n!a!b!}(x) = \left(\frac{1}{\pi}\right)^{3/2}\left(\frac{1}{2}\right)^{(n+a+b)/2}\left(\frac{1}{nab}\right)^{1/2} H_n(z')H_a(x')H_b(y')$$
$$\times e^{-\frac{1}{2}\left(t'^2+z'^2+x'^2+y'^2\right)},$$

(A.15)

with the eigenvalue $\lambda = n + a + b$.

The above oscillator wave functions are separable in the Cartesian coordinate system and the transverse coordinate variables are not affected by the Lorentz boost along the z-direction. Therefore the factors depending on the x- and y-variables can be omitted when studying the Lorentz transformation properties for these oscillator wave functions. The essential part of the covariant wave function is

$$\psi_\eta^n(t, z) = \left(\frac{1}{\pi}\right)^{1/2}\left(\frac{1}{2}\right)^{n/2}\left(\frac{1}{n!}\right)^{1/2} H_n(z')e^{-\frac{1}{2}\left(t'^2+z'^2\right)}.$$

(A.16)

We study next the orthogonality relations of the wave functions. As the volume element is Lorentz-invariant,

$$dzdt = dz'dt',$$

(A.17)

there is no difficulty in understanding the orthogonality relation:

$$\int \psi_\eta^n(t, z)\psi_\eta^m(t, z)dtdz = \int \psi_0^n(t, z)\psi_0^m(t, z)dtdz = \delta_{mn}.$$

(A.18)

The inner product of two wave functions belonging to different Lorentz frames is a more interesting problem. The orthogonality relation is [60]

$$\int \psi_\eta^n(t, z)\psi_0^m(t, z)dtdz = \left(1 - \beta^2\right)^{(n+1)/2}\delta_{nm}.$$

(A.19)

It is a remarkable fact that the orthogonality in the quantum number n is still preserved. This is a result of the Lorentz invariance of the harmonic oscillator differential equation. The oscillator equation does not depend on the velocity parameter β.

Examining the factor $(1 - \beta^2)^{(n+1)/2}$ in equation (A.19), we first note that, when the oscillator is in the ground state, it becomes like a Lorentz contraction of a rigid rod by $(1 - \beta^2)^{1/2}$. Then we can obtain excited-state wave functions from the ground state wave function through repeated applications of the step-up operator defined in equation (1.135):

$$\psi_\eta^n(x) = \left(\frac{1}{n!}\right)^{1/2} (a^\dagger)^n \psi_\eta^0(x).$$

(A.20)

Therefore, if the ground-state wave function is like a rigid rod along the z-direction, the nth excited state should behave like a multiplication of $(n + 1)$ rigid rods [43]. This contraction property is summarized in figure A.1.

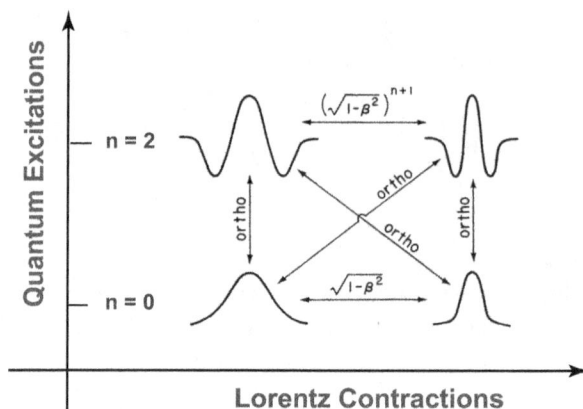

Figure A.1. Orthogonality and Lorentz contraction properties of the normalizable harmonic oscillator wave functions. The ground-state wave function is contracted like a rigid rod. The nth excited state is contracted like a product of $(n + 1)$ rigid rods [41, 43, 60].

Poincaré formulated the group theory of Lorentz transformations applicable to the four-dimensional Minkowskian space consisting of three space coordinates and one time variable. We use in this book, the four-vector convention (t, z, x, y). Poincaré started by considering the six generators performing three rotations and three Lorentz boosts. In chapter 2 this group was called the Lorentz group, but it is also known as the homogeneous Lorentz group.

Additionally, Poincaré considered translations applicable to those four space–time variables, with four generators. When these four generators are added to the six generators for the homogeneous Lorentz group, the result is the inhomogeneous Lorentz group [66] with ten generators. This larger group is called the Poincaré group in the literature, and the homogeneous Lorentz group is then a subgroup of this larger group.

The four translation generators produce space–time four-vectors consisting of energy and momentum. Thus, within the framework of the Poincaré group, Wigner [66] considered the subgroups of the homogeneous Lorentz group for a fixed value of momentum. As we saw in chapter 3, Wigner's little groups define the internal space–time symmetry of the particle.

Let us go back to the coordinates x_a and x_b in equation (A.2). These represented the quarks bound in a hadron. Now let us consider performing inhomogeneous Lorentz transformations on these quarks. The same Lorentz transformation matrix is applicable to x_a, x_b, x, and X. We find, however, that under the space–time translation which changes x_a and x_b to $x_a + a$ and $x_b + a$, although X becomes $X + a$, x remains invariant. Translations do not affect the quark separation coordinate x. For this reason, the generators of translations for this system are

$$P_\mu = -i\frac{\partial}{\partial X^\mu}.$$
(A.21)

The generators of homogeneous Lorentz transformations are [49]

$$M_{\mu\nu} = S_{\mu\nu} + L_{\mu\nu}, \tag{A.22}$$

where

$$S_{\mu\nu} = -i\left(X_\mu \frac{\partial}{\partial X^\nu} - X_\nu \frac{\partial}{\partial X^\mu}\right) \quad \text{and} \quad L_{\mu\nu} = -i\left(x_\mu \frac{\partial}{\partial x^\nu} - x_\nu \frac{\partial}{\partial x^\mu}\right). \tag{A.23}$$

Our interest now is in constructing normalizable wave functions which are diagonal in the Casimir operators P^2 and W^2 [43],

$$P^2 = \left(\frac{\partial}{\partial X^\mu}\right)^2 = \frac{1}{2}\left(\frac{\partial^2}{\partial x_\mu^2} - x_\mu^2\right) - m_0^2 \tag{A.24}$$

and

$$W^2 = M^2(\mathbf{L})^2, \tag{A.25}$$

where $\mathbf{L}$ is the operator in three dimensional space:

$$L_i = -i\varepsilon_{ijk}x_j \frac{\partial}{\partial x_k}.$$

We see that the eigenvalue of $-P^2$ is $M^2 = m_0^2 + (\lambda + 1)$, and that the eigenvalue for W^2 is $M^2\ell(\ell + 1)$. Here M is the hadronic mass, and ℓ is the total intrinsic angular momentum of the hadron due to internal motion of the spinless quarks [49].

Additionally, the solutions can be chosen to be diagonal in the component of the intrinsic angular momentum along the direction of the motion. We have previously called this the helicity and when the hadron moves along the Z-direction, the helicity operator is L_3.

We note that the spatial part of the harmonic oscillator equation in equation (A.10) is also separable in the spherical coordinate system. Hence, its solution using spherical variables in the hadronic rest frame space spanned by z', x', and y' can be written. In its most general form the solution is

$$\psi_{\eta\lambda\ell}^{k,\,m}(x) = R_\mu^\ell(r')Y_\ell^m(\theta',\,\phi')(1/\sqrt{\pi}2^k k!)^{1/2}H_k(t')e^{-t'^2/2}, \tag{A.26}$$

where

$$r' = (z'^2 + x'^2 + x'^2)^{1/2}, \quad \theta' = \cos^{-1}\left(\frac{z'}{r'}\right), \quad \text{and} \quad \phi' = \tan^{-1}\left(\frac{y'}{x'}\right)$$

and

$$\lambda = 2\mu + \ell - k. \tag{A.27}$$

Here $R_\mu^\ell(r')$ is the normalized radial wave function for the three-dimensional harmonic oscillator:

$$R_\mu^\ell(r) = \left(2(\mu!)/(\mu + \ell + 3/2)^3\right)^{1/2}r^\ell L_\mu^{\ell+1/2}(r^2)e^{-r^2/2}. \tag{A.28}$$

The associated Laguerre function is given by $L_\mu^{\ell+1/2}(r^2)$. The spherical form given in equation (A.26) can also be expressed as a linear combination of the wave functions in the Cartesian coordinate system given in equation (A.15).

We note that the wave function of equation (A.26) is diagonal in L_3 and also in the Casimir operators of equations (A.24) and (A.25). It does form a vector space for the $O(3)$-like little group, but the system is infinitely degenerate due to excitations along the t'-axis. We can suppress the time-like oscillation, as we did in section A.1.1, by imposing the subsidiary condition of equation (A.14). Then the solution becomes

$$\psi_{\eta\lambda\ell}^m(x) = R_\mu^\ell(r') Y_\ell^m(\theta', \phi') \{ (1/\pi)^{1/4} e^{-t'^2/2} \}, \qquad (A.29)$$

where $\lambda = 2\mu + \ell$. This means that for a given λ, there are only a finite number of solutions. Thus we can express the above spherical form as a linear combination of the solutions without time-like excitations in the Cartesian coordinate system given in equation (A.15).

This allows us to write the solution of the differential equation of (A.1) and (A.3) as

$$\phi(X, x) = e^{\pm iP \cdot X} \psi_{\eta\lambda\ell}^m(x). \qquad (A.30)$$

This wave function describes a free hadron with a definite four-momentum having an internal space–time structure which can be described by an irreducible unitary representation of the Poincaré group. The representation is unitary because the portion of the wave function depending on the internal variable x is square-integrable, and all the generators of Lorentz transformations are Hermitian operators.

A.1.3 Lorentz transformations of harmonic oscillator wave functions

As the physics of the covariant harmonic oscillator formalism is now understood, the discussion of its covariance properties can continue. We return to the covariant form given in equation (A.16).

The wave function becomes that of the hadron at rest when $\beta = 0$. As β increases, the wave function should be of the form

$$\psi_\eta^n(t, z) = \psi_\eta^{n,0}(t, z) = \sum_{m,k} A_{mk}^n(\beta) \psi_0^{m,k}(t, z), \qquad (A.31)$$

where

$$\psi_\eta^{m,k}(t, z) = \left(\frac{1}{\pi}\right)^{1/2} \left(\frac{1}{2}\right)^{(m+k)/2} \left(\frac{1}{m!k!}\right)^{1/2} H_k(t') H_m(z') e^{-\frac{1}{2}(t'^2 + z'^2)}. \qquad (A.32)$$

This is a linear combination of the wave functions in the rest frame. Based on the conclusion of section A.1.2 we know that there are no time-like oscillations, and therefore that the physical wave functions should have $k = 0$. The left-hand side is a physical wave function. This should not keep us from expanding equation (A.31) into a complete set of orthonormal wave functions which include time-like excitations. The summation may still result is a physical wave function, even though each term on the right-hand side may not be a physical wave function.

The oscillator differential equation is Lorentz invariant, and thus the eigenvalue n for $\psi_\eta^n(t, z)$ of equation (A.31) remains invariant. Therefore only the terms which satisfy the condition

$$n = (m - k) \tag{A.33}$$

make non-zero contributions in the sum. This allows us to simplify the above expression to

$$\psi_\eta^n(t, z) = \sum_k A_k^n(\beta)\psi_0^{n+k,k}(t, z). \tag{A.34}$$

The above wave function is an infinite-dimensional unitary representation of the Lorentz group [48], but it only depends on one variable k.

Now the coefficient $A_k^n(\beta)$ must be determined. If we use the orthogonality relation of the Hermite polynomials, we can write

$$A_k^n(\beta) = \frac{1}{\pi}\left(\frac{1}{2}\right)^{(n+k)}\left(\frac{1}{n!(n+k)!k!}\right)^{1/2} \int H_{n+k}(z)H_k(t)H_n(z') \tag{A.35}$$
$$\times e^{-\frac{1}{2}\left(t^2+t'^2+z^2+z'^2\right)}dzdt.$$

The Hermite polynomials and the Gaussian form are mixed here with the kinematics of Lorentz transformation. If we use the generating function for the Hermite polynomial, the evaluation of the integral is straightforward [43, 60], and we obtain the result

$$A_k^n(\beta) = (1 - \beta^2)^{(n+1)/2}\left(\frac{(n+k)!}{n!k!}\right)^{1/2}\beta^k. \tag{A.36}$$

As the binomial expansion of $(1 - \beta^2)^{-(n+1)}$ is

$$(1 - \beta^2)^{-(n+1)} = \sum_k \frac{(n+k)!}{n!k!}\beta^{2k}, \tag{A.37}$$

we are led to the result

$$\sum_k |A_k^n(\beta)|^2 = 1. \tag{A.38}$$

This relation confirms that the expansion given in equation (A.34) is a unitary transformation.

When $n = 0$, the transformation corresponds to the Lorentz boost of the ground-state wave function which is of the Gaussian form. If we go back to section 5.6 where the density matrix for thermally excited states was discussed, we note that the thermal excitation of the oscillator shares the same mathematics as the Lorentz boost of the ground-state wave function.

We can consider next the Lorentz-boosted harmonic oscillator wave function in the light-cone coordinate system. As was noted in section 3.2, the basic advantage of the light-cone coordinate system is that the Lorentz boost can be formulated in terms of a squeeze in the zt-plane. For the covariant harmonic oscillators, it is the Gaussian factor which determines the localization property of the wave function. The Gaussian factor is localized within the region

$$(t'^2 + z'^2) < 1. \tag{A.39}$$

When the hadron is at rest, this becomes a circular region in the zt-plane with $(t^2 + z^2) < 1$. In terms of the light-cone variables, as defined in equation (3.28) this circular region is

$$(z_+^2 + z_-^2) < 1. \tag{A.40}$$

When the hadron moves, the region becomes

$$\left(\frac{1+\beta}{1-\beta}\right)z_+^2 + \left(\frac{1-\beta}{1+\beta}\right)z_-^2 < 1. \tag{A.41}$$

Certainly this is a squeezed circle. In terms of the zt variables, this elliptic region can be expressed as

$$\frac{1}{2}\left(\frac{1+\beta}{1-\beta}\right)(t+z)^2 + \frac{1}{2}\left(\frac{1-\beta}{1+\beta}\right)(t-z)^2 < 1. \tag{A.42}$$

We can then conclude that the Lorentz boost on the harmonic oscillator wave function results in a squeeze in the zt-plane. From section 3.2 we know that the squeeze property is applicable to the energy–momentum plane. The squeeze property of the covariant oscillators in phase space will be examined in section A.1.4.

A.1.4 Covariant phase-space picture of harmonic oscillators

Let us now examine the covariant harmonic oscillator formalism using the phase-space picture of quantum mechanics. As we have seen here, the harmonic oscillator wave function consists of a Gaussian factor and Hermite polynomials.

Because the localization property of the wave function is determined by the Gaussian factor, we study first the ground-state wave function, whose form is

$$\psi_0^0(t, z) = \left(\frac{1}{\pi}\right)^{1/2} e^{-(t^2+z^2)/2}. \tag{A.43}$$

In considering the excited states, we see that there are no time-like oscillations in the hadronic rest frame, and that the oscillations in the transverse directions are not affected. The only factor to be considered is the Hermite polynomial $H_n(z')$ multiplied by the ground-state wave function.

The pure-state density matrix for the covariant oscillator wave functions defined in equation (A.16) is

$$\rho_\eta^n(z, t; z', t') = \psi_\eta^n(z, t)\psi_\eta^{n*}(z', t').$$ (A.44)

This function satisfies the condition $\rho^2 = \rho$:

$$\rho_\eta^n(z, t; x', t') = \int \rho_\eta^n(z, t; x'', t'')\rho_\eta^n(z'', t''; z', t')dz''dt''.$$ (A.45)

It is not possible, in the present form of quantum mechanics, to take into account the time separation variables. To overcome this we take the trace of the matrix with respect to the t variable with the resulting density matrix

$$\rho_\eta^n(z, z') = \int \psi_\eta^n(z, t)\psi_\eta^{n*}(z', t)dt$$
$$= \left(\frac{1}{\cosh \eta}\right)^{2(n+1)} \sum_k \frac{(n + k)!}{n!k!}(\tanh \eta)^{2k}\psi_{n+k}(z)\psi_{n+k}^*(z').$$ (A.46)

Here,

$$\frac{1}{\cosh \eta} = \sqrt{1 - \beta^2} \quad \text{and} \quad \tanh \eta = \frac{v}{c} = \beta.$$ (A.47)

Although this density matrix has a trace of one, the trace of ρ^2 is less than one, as can be seen from

$$\text{Tr}(\rho^2) = \int \rho_\eta^n(z, z')\rho_\eta^n(z', z)dzdz'$$
$$= \left(\frac{1}{\cosh \eta}\right)^{4(n+1)} \sum_k \left[\frac{(n + k)!}{n!k!}\right]^2 (\tanh \eta)^{4k}.$$ (A.48)

This is less than one and is the result of our not knowing how, in the present formulation of quantum mechanics, to deal with the time-like separation of the quarks.

Calculating the entropy defined as

$$S = -\text{Tr}(\rho \ln(\rho)),$$ (A.49)

is the standard way to measure this ignorance. By pretending to know the distribution along the time-like direction and using the pure-state density matrix given in equation (A.44) we find that the entropy is zero. If, however, it is not known how to deal with the distribution along t, then the density matrix of equation (A.46) should be used to calculate the entropy. The resulting entropy value is

$$S = (n + 1)\{(\cosh \eta)^2 \ln(\cosh \eta)^2 - (\sinh \eta)^2 \ln(\sinh \eta)^2\}$$
$$- \left(\frac{1}{\cosh \eta}\right)^{2(n+1)} \sum_k \frac{(n + k)!}{n!k!} \ln\left[\frac{(n + k)!}{n!k!}\right](\tanh \eta)^{2k}.$$ (A.50)

We can write this entropy in terms of the velocity parameter β of the hadron,

$$
\begin{aligned}
S = &-(n+1)\left\{\ln(1-\beta^2) + \frac{\beta^2(\ln\beta^2)}{(1-\beta^2)}\right\} \\
&-(1-\beta^2)^{(n+1)} \sum_k \frac{(n+k)!}{n!k!} \ln\left[\frac{(n+k)!}{n!k!}\right]\beta^{2k}.
\end{aligned}
\tag{A.51}
$$

It is possible to expand the wave function given equation (A.16) in terms of the z and t variables [43, 50, 59, 60], and the result is

$$
\psi_\eta^n(z,t) = \left(\frac{1}{\cosh\eta}\right)^{(n+1)} \sum_k \left[\frac{(n+k)!}{n!k!}\right]^{1/2} (\tanh\eta)^k \chi_{n+k}(z)\chi_k(t),
\tag{A.52}
$$

where $\chi_n(z)$ is the nth excited-state oscillator wave function which takes the familiar form

$$
\chi_n(z) = \left[\frac{1}{\sqrt{\pi}2^n n!}\right]^{1/2} H_n(z)e^{-z^2/2}.
\tag{A.53}
$$

If $n = 0$, this formula becomes simplified to [60]

$$
\psi_\eta^0(z,t) = \left(\frac{1}{\cosh\eta}\right)^{1/2} \sum_k (\tanh\eta)^k \chi_k(z)\chi_k(t).
\tag{A.54}
$$

This formula also plays an important role in squeezed states of light [68] and also in continuous-variable entanglement [1, 6, 8, 14, 24, 58, 67].

The localization property of the wave function given in equation (A.52) is dictated by the Gaussian factor, which corresponds to the ground-state wave function. This is illustrated in figure A.2. We thus expect that much of the behavior of the density matrix or the entropy for the nth excited state to be the same as that for the ground state with $n = 0$.

The density matrix for the ground state with $n = 0$ can be computed from the Gaussian form of equation (A.54), and becomes

$$
\rho(z,z') = \left(\frac{1}{\pi\cosh(2\eta)}\right)^{1/2} \exp\left\{-\frac{1}{4}\left[\frac{(z+z')^2}{\cosh(2\eta)} + (z-z')^2\cosh(2\eta)\right]\right\}.
\tag{A.55}
$$

The entropy for this ground state then becomes

$$
S = (\cosh\eta)^2 \ln(\cosh\eta)^2 - (\sinh\eta)^2 \ln(\sinh\eta)^2.
\tag{A.56}
$$

This entropy can then be written in terms of the velocity parameter β as

$$
S = -\left\{\ln(1-\beta^2) + \frac{\beta^2(\ln\beta^2)}{(1-\beta^2)}\right\}.
\tag{A.57}
$$

This expression is consistent with the harmonic oscillator in thermal equilibrium if β^2 is identified with the Boltzmann factor $\exp(-\hbar\omega/k_BT)$, as discussed in section 8.4.

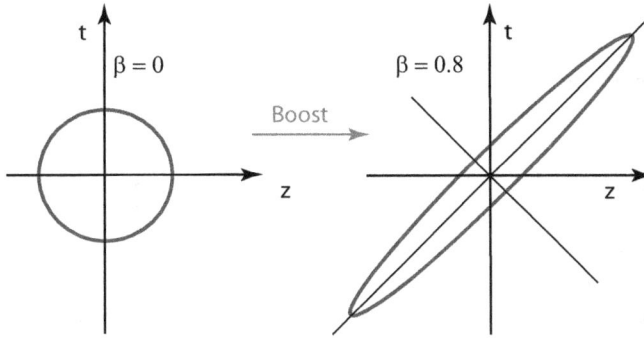

Figure A.2. Lorentz-squeezed quantum mechanics. Dirac's attempt for relativistic quantum mechanics starts from the Gaussian distribution. This figure shows how the Gaussian distribution appears to an observer moving with the velocity parameter $\beta = v/c = 0.8$ [49].

From equation (A.55) the width of the distribution is given by $\sqrt{\cosh(2\eta)}$. If this is written in terms of β,

$$\sqrt{\cosh(2\eta)} = \sqrt{\frac{1 + \beta^2}{1 - \beta^2}}, \tag{A.58}$$

it can be seen that this becomes very large as $\beta \to 1$.

Although the time separation variable exists in the Lorentz-covariant world, we pretend not to know about it. By not allowing measurement of this time separation, it becomes translated into the entropy.

The uncertainty in the measurement process can be seen from the simplest form of the time-independent Wigner function defined in equation (5.6). We write it here for the z and p variables as [44, 65]

$$W(z, p) = \frac{1}{\pi} \int \rho(z + y, z - y)e^{2ipy}dy. \tag{A.59}$$

When this Wigner function is integrated it becomes

$$W(z, p) = \frac{1}{\pi \cosh(2\eta)} \exp\left\{-\left(\frac{z^2 + p^2}{\cosh(2\eta)}\right)\right\}. \tag{A.60}$$

Here p designates the momentum conjugate to z not the hadronic momentum.

A.2 Quark–parton puzzle

Since the proton now consists of quarks distributed within a finite space–time region, the scattering amplitude will depend on the way in which quarks are distributed within the proton. In the context of high-energy electron–proton scattering experiments, an electron emits a virtual photon, which then interacts with the proton. Like the hydrogen atom, the proton has a localized probability distribution. The portion of the scattering amplitude which describes the interaction between the virtual photon and the proton is called the form factor. We expect the wave function in the quark model will describe the charge distribution. Since we are dealing with high-energy experiments with high momentum transfer, it is important to know how to describe the quark-model wave functions for rapidly moving protons.

The effectiveness of Yukawa's harmonic oscillator wave function in the relativistic quark model was first demonstrated by Fujimura *et al* [22] who showed that the Yukawa wave function leads to the correct high-energy asymptotic behavior of the nucleon form factor (see section A.2.3). Feynman *et al* [20] advocated the use of relativistic harmonic oscillators instead of Feynman diagrams for studying hadronic structures and interactions. Although the paper by Feynman *et al* contains all the troubles expected from relativistic wave equations, the authors did not make any attempt to hide those troubles or suggest a solution.

A.2.1 Lorentz-covariant quark model

In terms of the quark model, hadrons are all bound states of quarks. In this model, baryons are bound states of three quarks and mesons are bound states of one quark and one anti-quark. That the hadronic mass spectra are like those of three-dimensional harmonic oscillators was shown by Feynman *et al* [20]. Among the early successes of the quark model was the successful calculation of the ratio of the proton–neutron electromagnetic potential and magnetic moments [4].

We could ask how the mass spectrum calculated within the framework of non-relativistic quantum mechanics is valid for this relativistic case, since the time separation variable is ignored. The answer given by Feynman *et al* is not satisfactory. The correct answer is that for massive particles Wigner's little group is like the three-dimensional rotation group. With regard to the role of the time separation variable, this issue was discussed in chapter 3 and in section A.1.

The original question which might have occurred in discussions between Bohr and Einstein might have been how the hydrogen atom looks to a moving observer. It was not possible then, and is still not possible today, to accelerate the hydrogen atom to speeds close to that of light. Now, however, we can learn about this by studying the proton in the quark model based on the three-dimensional harmonic oscillator. For the hydrogen atom, the Coulomb potential is used, while the harmonic oscillator potential provides the binding force between quarks. These two different potentials, however, share the same quantum mechanics.

What is needed for this purpose is a bound-state wave function which can be Lorentz-boosted. The harmonic oscillator wave function discussed in section A.1 provides the natural choice. Recall that the ground-state wave function can be Lorentz boosted and that the harmonic oscillator wave function is separable in the Cartesian coordinate system. Therefore, we consider only the longitudinal and time-like coordinates and leave out the transverse components of the wave function. For this purpose, let us rewrite the wave function of equation (A.6) as

$$\frac{1}{2}\left[-\left(\frac{\partial}{\partial x}\right)^2 + x^2\right]\psi(x) = (\lambda + 1)\psi(x). \tag{A.61}$$

If now we Lorentz boost equation (A.61) and drop the transverse coordinates we obtain

$$\frac{1}{2}\left\{\left[z^2 - \left(\frac{\partial}{\partial z}\right)^2\right] - \left[t^2 - \left(\frac{\partial}{\partial t}\right)^2\right]\right\}\psi(z, t) = \lambda\psi(z, t). \tag{A.62}$$

Remembering that there are no time-like excitations, the solution to this equation becomes

$$\psi_0^n(z, t) = \chi_n(z)\chi_0(t) = \left[\frac{1}{\pi 2^n n!}\right]^{1/2} H_n(z)\exp\left[-\frac{1}{2}(z^2 + t^2)\right], \tag{A.63}$$

where $H_n(z)$ is the Hermite polynomial of order n, and $\chi_n(z)$ is the oscillator wave function in the nth excited state:

$$\chi_n(z) = \left[\frac{1}{\sqrt{\pi}2^n n!}\right]^{1/2} H_n(z)\exp\left(-\frac{1}{2}z^2\right). \tag{A.64}$$

The differential equation of (A.62) is invariant under the Lorentz boost along the z-direction. The Lorentz boost of the wave function is obtained by writing

$$\psi_\eta^n(z, t) = \psi_0^n(z', t') = \left[\frac{1}{\pi 2^n n!}\right]^{1/2} H_n(z')\exp\left[-\frac{1}{2}(z'^2 + t'^2)\right]. \tag{A.65}$$

For $n = 0$, the wave function has the Gaussian form

$$\psi_\eta^0(z, t) = \left(\frac{1}{\pi}\right)^{1/2} \exp\left\{-\frac{1}{4}\left[e^{-2\eta}(z + t)^2 + e^{2\eta}(t - z)^2\right]\right\}, \tag{A.66}$$

which becomes

$$\psi_0(z, t) = \frac{1}{\sqrt{\pi}} \exp\left\{-\frac{1}{2}\left[(z^2 + t^2)\right]\right\}, \tag{A.67}$$

for $\eta = 0$.

We can now consider the overlap of the wave function $\psi_\eta^n(z, t)$ with that with $\eta = 0$. This is shown in figure A.3. Of interest is the integral

$$\int \left(\psi_\eta^{n'}(z, t)\right)^* \psi_0^n(z, t)\,dz\,dt, \tag{A.68}$$

where $\psi_0^n(z, t)$ is given by

$$\psi_0^n(z, t) = \left[\frac{1}{\pi 2^n n!}\right]^{1/2} H_n(z)\exp\left[-\frac{1}{2}(z^2 + t^2)\right]. \tag{A.69}$$

Here $H_n(z)$ is the Hermite polynomial of order n and ψ_0^n and can be written as

$$\psi_0^n = \chi_n(z)\chi_0(t). \tag{A.70}$$

A-16

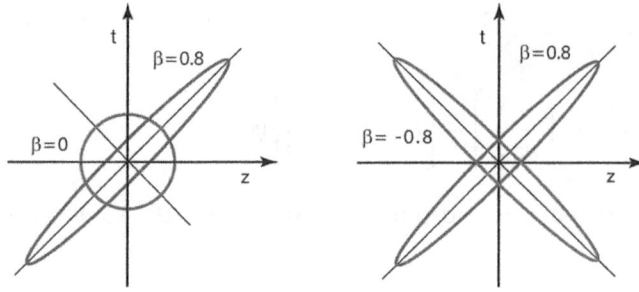

Figure A.3. Two overlapping wave functions. The wave functions with $\beta = 0$ and $\beta = 0.08$ are sketched in figure A.2. They can overlap as shown in the present figure. The wave functions can also move in the opposite directions with $\beta = \pm 0.8$ [3].

The result of this integral is [60]

$$\left(\psi_\eta^{n'}, \psi_0^n\right) = \left(\frac{1}{\cosh \eta}\right)^{(n+1)} \delta_{nn'}. \tag{A.71}$$

As we saw in section A.1, the orthogonality relation is preserved between two wave functions in two different frames.

The contraction given on the right-hand side of equation (A.71) results from the inner product between two wave functions if $n = n'$. From equation (A.47) we can express this in terms of the velocity parameter $\beta = v/c$, where v is the hadronic velocity. Thus we obtain

$$\frac{1}{\cosh \eta} = \sqrt{1 - \beta^2}. \tag{A.72}$$

As this expression is more familiar to us, the right-hand side of equation (A.71) can be written as

$$\left(\sqrt{1 - \beta^2}\right)^{(n+1)} \delta_{nn'}. \tag{A.73}$$

As we saw in section A.1, the ground-state wave function with $n = 0$ is like the Lorentz contraction of a rigid rod. The first excited state is like an additional rod. This is reasonable in view of the fact that the excited states are obtained through application of the step-up operator. The nth excited state $|n\rangle$ can be written as

$$\frac{1}{\sqrt{n!}}(a^\dagger)^n|0\rangle, \tag{A.74}$$

as we saw in equation (A.20). The step-up operator is responsible for the additional contraction factor $\sqrt{1 - \beta^2}$.

If we replace the value of η in one of the wave functions by the non-zero value η', $\cosh \eta$ in equation (A.71) should become $\cosh(\eta - \eta')$. An interesting possibility to consider is when $\eta' = -\eta$, as shown in figure A.3. This is then an overlap of two wave functions moving in opposite directions, and the contraction factor is

$$\left(\frac{1}{\cosh(2\eta)}\right)^{(n+1)}.$$

(A.75)

In terms of β this becomes

$$\left(\frac{1 - \beta^2}{1 + \beta^2}\right)^{(n+1)}.$$

(A.76)

It appears possible now, based on these orthogonality and contraction properties, to give a quantum probability interpretation to the covariant harmonic oscillator in the Lorentz-covariant world, as shown in figure A.1. These wave functions, however, contain the time separation variable between the constituent particles.

A.2.2 Feynman's parton picture

In discussing what happens to the wave function when the proton is Lorentz-boosted, we need to examine the two-body problem. In section A.1.3 we have discussed the Lorentz-squeeze problem. We said above that hadrons are quantum bound states of quarks having a localized probability distribution. It is the localization condition that in all bound states is responsible for the existence of discrete mass spectra. The most convincing evidence for this bound-state picture is the hadronic mass spectra [20, 43].

Unfortunately this bound-state picture is applicable only to observers in the Lorentz frame in which the hadron is at rest. We need to ask how the hadrons appear to observers in other Lorentz frames.

It was Feynman who in 1969 observed that a fast-moving hadron can be regarded as a collection of many *partons*. These partons appear to have properties which are quite different from those of the quarks [17, 18, 43]. These properties include that the number of quarks inside a static proton is three, while the number of partons in a rapidly moving proton appears to be infinite. How can it be that a proton looking like a bound state of quarks to one observer can appear so differently to an observer in a different Lorentz frame? The following systematic observations were made by Feynman:

1. The picture is valid only for hadrons moving with velocity close to that of light.
2. The interaction time between the quarks becomes dilated, and partons behave as free independent particles.
3. The momentum distribution of partons becomes wide-spread as the hadron moves fast.
4. The number of partons seems to be infinite or much larger than that of quarks.

If the hadron is believed to be a bound state of two or three quarks, each of the above phenomena appears as a paradox, particularly 2 and 3 together. How can a free particle have a wide-spread momentum distribution?

To resolve this paradox, we construct the momentum–energy wave function corresponding to equation (A.66). If the quarks have the four-momenta p_a and p_b, we can construct two independent four-momentum variables [20]

$$P = p_a + p_b \quad \text{and} \quad q = \sqrt{2}\,(p_a - p_b). \tag{A.77}$$

The four-momentum P is the total four-momentum and is thus the hadronic four-momentum while q measures the four-momentum separation between the quarks.

This results in a momentum–energy wave function

$$\phi_\eta(q_z, q_0) = \left(\frac{1}{\pi}\right)^{1/2} \exp\left\{ -\frac{1}{4}\left[e^{-2\eta}(q_0 + q_z)^2 + e^{2\eta}(q_0 - q_z)^2 \right] \right\}. \tag{A.78}$$

When η becomes large, $q_0 = q_z$, and the wave function becomes

$$\phi_\eta(q_z) = \left(\frac{1}{\pi}\right)^{1/4} \exp\left\{ -\left[e^{-2\eta}(q_z)^2 \right] \right\}. \tag{A.79}$$

Since we are using the harmonic oscillator, the mathematical form of the equation (A.78) is identical to that of the space–time wave function of equation (A.66). Also the same are the Lorentz-squeeze properties of these wave functions. This aspect of the squeeze is illustrated in figure A.4 and has been exhaustively discussed in the literature [36, 51]. Furthermore, the hadronic structure function calculated from this formalism is in reasonable agreement with experimental data [31].

If the hadron is at rest with $\eta = 0$, the behavior of both wave functions is like that for the static bound state of quarks. As η increases, the wave functions become continuously squeezed until they become concentrated along their respective positive

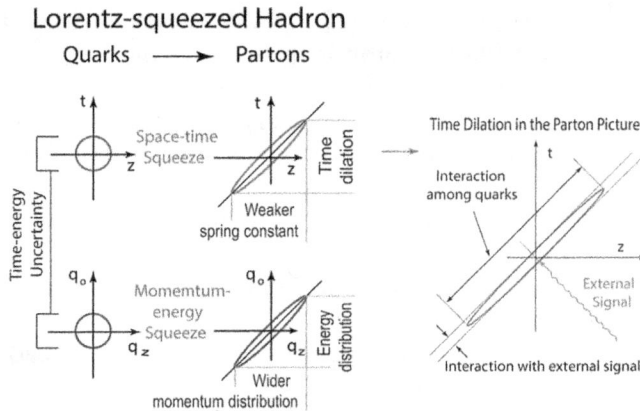

Figure A.4. Lorentz-squeezed space–time and momentum–energy wave functions. As the hadron's speed approaches that of light, both wave functions become concentrated along their respective positive light-cone axes. These light-cone concentrations lead to Feynman's parton picture [36, 51]. The external signal, since it is moving in the direction opposite to the direction of the hadron, travels along the negative light-cone axis. Thus, the interaction time of this signal with the bound state is much shorter than the period of oscillation of the quarks inside the hadron. This effect is called Feynman's time dilation [5, 17, 18, 45].

light-cone axes. If we look at the z-axis projection of the space–time wave function, we see that the width of the quark distribution increases as the hadronic speed approaches that of the speed of light. It is then clear that the position of each quark appears wide-spread to the observer in the laboratory frame, and the quarks appear like free particles.

The space–time wave function and the momentum–energy wave function are alike. That the longitudinal momentum distribution becomes wide-spread as the hadronic speed approaches the velocity of light contradicts what we expect from non-relativistic quantum mechanics which is that the width of the momentum distribution is inversely proportional to that of the position wave function. We expect that if quarks are free, they must have a sharply defined momenta, not a wide-spread distribution.

According to our Lorentz-squeezed space–time and momentum–energy wave functions, however, the space–time width and the momentum–energy width increase in the same direction as the hadron is boosted. This is an effect of Lorentz covariance and leads to the resolution of one of the quark–parton puzzles [36, 43, 51].

Another aspect of the parton picture puzzle is that partons appear as incoherent particles, while quarks are coherent when the hadron is at rest. Is coherence destroyed by the Lorentz boost [37, 38]? The answer is no, and the resolution to the puzzle is as follows.

The hadronic matter, when the hadron is Lorentz-boosted, becomes squeezed and concentrated in the elliptic region along the positive light-cone axis. The length of the major axis thus expands by e^{η}, and the minor axis contracts by $e^{-\eta}$.

This results in the interaction time of the quarks among themselves becoming dilated. As the wave function becomes wide-spread, the distance increases between one end of the oscillator well and the other end. This effect, universally observed in high-energy hadronic experiments, was first noted by Feynman [17, 18]. Although the period of oscillation increases like e^{η}, the external signal, which is moving in the direction opposite to the that of the hadron, travels along the negative light-cone axis.

When the hadron contracts along the negative light-cone axis, the interaction time decreases by $e^{-\eta}$ causing the ratio of the interaction time to the oscillator period to become $e^{-2\eta}$. Since the energy of each proton coming out of the LHC accelerator is 13 TeV, the ratio becomes 1.25×10^{-9}, which is indeed a small number. The external signal cannot sense the interaction of the quarks among themselves inside the hadron.

We see therefore, that the covariant harmonic oscillator formalism provides one Lorentz-covariant entity which produces the quark and parton models as two limiting phenomenon. This is indicated in table A.1.

A.2.3 Proton structure function and form factor

The quark distribution as measured in momentum–energy space is called the proton structure function. It can be measured from the inelastic electron–proton scattering

Table A.1. The unified picture of the quark and parton models can be viewed as further content of Einstein's energy–momentum relation [36, 43].

	Massive slow	Lorentz covariance	Massless fast
Energy–momentum	$E = p^2/2m$	Einstein's $E = [p^2 + m^2]^{1/2}$	$E = p$
Spin, helicity, gauge trans.	J_3, J_1, J_2	Wigner's little group	J_3, N_1, N_2
Hadron's constituents	Gell-Mann's quark model	Lorentz-covariant harmonic oscillator	Feynman's parton picture

with one-photon exchange [5]. It is now of interest to see how close the Gaussian form of equation (A.78) is to the experimental world.

We saw previously that in the large-η limit, the proton wave function is within the narrow elliptic region where $q_z = q_0$. Thus we are left with the wave function depending on only one variable as was given in equation (A.79). If we use equation (A.77), we can write the momentum of each quark as

$$p_{az} = \left(\frac{P_z}{2} + \frac{q_z}{2\sqrt{2}} \right) \quad \text{and} \quad p_{bz} = \left(\frac{P_z}{2} - \frac{q_z}{2\sqrt{2}} \right). \tag{A.80}$$

Because the external signal only interacts with one quark, we can introduce the parameter

$$x = \frac{p_{az}}{P_z}, \tag{A.81}$$

which is the ratio of the quark momentum to the hadronic momentum. This is the variable used for measuring the parton distribution in high-energy laboratories.

We can now write the Gaussian form of equation (A.79) in terms of this x variable, and thus write the quark distribution as

$$d(x) = \exp\left[-\gamma \left(x - \frac{1}{2} \right)^2 \right]. \tag{A.82}$$

Here the constant γ is to be determined from the level separation of the hadronic mass spectra [20]. The variable x can range from a minimum value of zero to a maximum value of 1 with a peak in the Gaussian form at $x = 1/2$.

The above discussion is suitable for a hadron consisting of two quarks. However, in order to make contact with the experimental world, we have to take into account that the proton is a bound state of three quarks. Fortunately, within the formalism of the harmonic oscillator the three-body bound system can be separated into two independent oscillators. Feynman *et al* worked out this problem in their 1971 paper [20]. Here we will reproduce their calculation.

We can let x_a, x_b, x_c represent the space–time coordinates for the three quarks. Considering that there is an oscillator force between each pair of quarks, we obtain the quadratic form

$$[(x_a - x_b)^2 + (x_b - x_c)^2 + (x_c - x_a)^2]. \tag{A.83}$$

To deal more efficiently with this expression, Feynman *et al* used the following three variables:

$$X = \frac{x_a + x_b + x_c}{3},$$
$$r = \frac{x_a + x_b - 2x_c}{6}, \tag{A.84}$$
$$s = \frac{x_b - x_a}{2}.$$

These can be rewritten as

$$x_a = X - 2r,$$
$$x_b = X + r - \sqrt{3}\,s, \tag{A.85}$$
$$x_c = X + r + \sqrt{3}\,s.$$

Using only the r and s variables, we obtain the quadratic form

$$18(r^2 + s^2). \tag{A.86}$$

This does not depend on the X variable, which specifies the space–time coordinate of the proton.

Similarly for the momentum–energy four-vectors, we will use p_a, p_b, and p_c for the quarks a, b, c, respectively. We can then introduce the following variables for the momentum–energy four-vectors:

$$P = p_a + p_b + p_c,$$
$$q = p_a + p_b - 2p_c, \tag{A.87}$$
$$k = \sqrt{3}\,(p_b - p_a).$$

We then arrive at

$$p_a = \frac{1}{3}P + \frac{1}{6}q - \frac{1}{2\sqrt{3}}k,$$
$$p_b = \frac{1}{3}P + \frac{1}{6}q + \frac{1}{2\sqrt{3}}k, \tag{A.88}$$
$$p_c = \frac{1}{3}P - \frac{1}{3}q.$$

Again using only the q and k variables, we obtain the quadratic form

$$18(q^2 + k^2). \tag{A.89}$$

This does not depend on the P variable, which is a measure of the momentum and energy of the proton.

Let us suppose that the external signal interacts with quark c, then its momentum depends only on the q variable, which we can write as

$$q = P - 3p_c. \tag{A.90}$$

Once again it is possible to use the x variable as

$$x = \frac{p_{cz}}{P_z}. \tag{A.91}$$

The quark distribution should now take the form

$$d(x) = \left(\frac{1}{\pi\gamma}\right)^{1/2} \exp\left[-\gamma\left(x - \frac{1}{3}\right)^2\right]. \tag{A.92}$$

Here again the constant γ is determined from the hadronic mass spectra based on the oscillator model [20]. Figure A.5 compares this Gaussian form with what we observe in the real world as measured from inelastic electron–proton scattering [5]. The distributions shown in these two curves are somewhat different because the quarks do not interact with the incoming photon as a point particle.

To explain this difference, the simplest model would be to put all the effects into one additional quark in the oscillator system. This, however, makes the proton a bound state of four quarks. In this model it is the fourth quark which is responsible for all the unexplained effects [42]. Another model is known as the valon model [33, 34] which allows all those non-point effects to be eliminated. It was using this valon model that Hussar [31] derived the experimental curve which is compared with the Gaussian form as shown in figure A.5.

Even though this graph may not be as accurate as desired, the most remarkable feature is that the Gaussian form and the constant γ were calculated from the proton at rest. The constant γ came from the level spacing in the hadronic mass spectra. It is to be noticed that these two features manifest themselves for a proton whose speed is very close to that of the light.

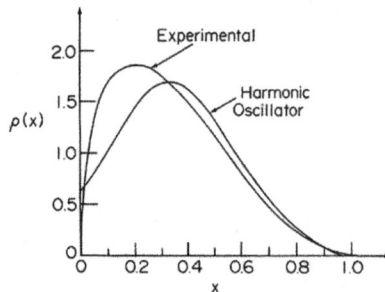

Figure A.5. Parton distribution function compared with experimental data. The boosted oscillator has its peak at $x = 1/3$. This Gaussian form gives a reasonable agreement with experimental data for large values of x, but the disagreement is substantial for small values of x. Reprinted figure with permission from [31], © 1981 by the American Physical Society.

Many other models have been proposed to deal with the problem of providing corrections to the parton distribution. Quantum chromodynamics (QCD) is one of the most prominent [7]. Although QCD provides corrections to the distribution, it does not produce the distribution from which to start. It is the covariant harmonic oscillator function which provides this starting point.

A similar situation occurs with respect to quantum electrodynamics (QED). Although QED was quite successful in producing the Lamb shift in the hydrogen energy spectrum, the Rydberg energy levels to which the correction is made cannot be produced. It is still necessary to obtain the hydrogen energy levels from the Schrödinger or Dirac equation with the localization condition on wave functions.

We will consider next the elastic scattering of a proton and an electron with the exchange of one photon. Only if the proton is a point particle, can the scattering cross section be calculated in a straightforward manner from the one-photon exchange Feynman diagram. If the proton's recoil is negligible, this process known as Rutherford scattering, results in the cross section becoming the same as the classical Coulomb scattering.

It was observed first by Hofstadter and McAllister in 1955 [30] that as the momentum transfer becomes substantial the cross section deviates from that of Rutherford scattering. This is indicated in figure A.6. It was subsequently observed that the cross section decreases as

$$\frac{1}{(\text{momentum transfer})^8}. \tag{A.93}$$

The fact that the proton is not a point particle and that the electric charge inside the proton is distributed with a finite radius is the cause of this deviation. The portion of the scattering amplitude describing the distribution of the electric charge inside the proton is called the proton form factor. Consequently, the proton form factor should therefore decrease as

$$\frac{1}{(\text{momentum transfer})^4}. \tag{A.94}$$

This decrease in the behavior of the form factor is known as the dipole cut-off in the literature. One of the major branches of high-energy physics results from this dipole cut-off and possible deviations from it. In the past there have been some far-reaching theoretical models to deal with this problem [21].

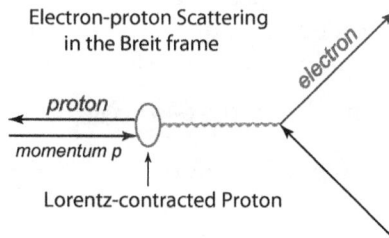

Figure A.6. Breit frame. The incoming and outgoing protons move with equal magnitude of momentum in opposite directions [43].

Here we will approach this problem using the harmonic oscillator formalism developed in section A.1. It will be shown that a consequence of the coherence between the contraction of the proton wave function and the decrease in the wavelength of the incoming signal is what produces the dipole cut-off.

The formalism of section A.1 is based mainly on the papers written by Dirac and Wigner. It is, however, interesting to note that the same harmonic oscillator functions can be derived from those authors who attempted to understand the proton form factor. These authors were not aware of the works of Dirac and Wigner. We briefly review what they did.

In 1953, Yukawa constructed a Lorentz-invariant differential equation. He achieved this by using harmonic oscillator wave functions that can be Lorentz-transformed [69]. Yukawa was primarily interested in the mass spectrum produced by his equation. Unfortunately, his mass spectrum did not appear to have anything to do with the physical world at that time.

In 1956, as the non-zero charge radius of the proton was observed [29, 30], Markov considered using Yukawa's oscillator formalism for calculating the proton form factor [55]. Unfortunately, the constituent particles of the oscillator wave functions were not defined at that time. Ginzburg and Man'ko [25], after the emergence of the quark model in 1964 [23], considered relativistic harmonic oscillators for bound states of quarks.

Fujimura, Kobayashi, and Namiki, without mentioning Yukawa's 1953 paper, used the quark model based on Yukawa's relativistic oscillator wave function to calculate the proton form factor, and obtained the dipole cut-off [22]. That same year, Licht and Pagnamenta [53] derived the same result using Lorentz-contracted oscillator wave functions. To by-pass the time separation variable appearing in the covariant formalism they used the Breit coordinate system [40].

Confirming the earlier suggestion made by Yukawa in 1953, Feynman et al [20], in 1971, noted that the observed hadronic mass spectra can be explained in terms of the degeneracies of three-dimensional harmonic oscillators. Although they quoted the paper by Fujimura et al [22], they did not mention Yukawa's 1953 paper [69]. Feynman et al could not write down normalizable wave functions, and not taking into account Yukawa's work is perhaps the reason why.

Returning to the formalism developed in section A.1, we can consider that in the scattering of one electron and one proton by exchanging one photon it is possible to choose the Lorentz frame in which the incoming and outgoing protons are moving in opposite directions with the same speed. If we assume that the proton is moving along the z-direction as indicated in figure A.6, and that p is the magnitude of the momentum, the initial and final momentum–energy four-vectors are

$$(p, E) \quad \text{and} \quad (-p, E), \tag{A.95}$$

respectively, where $E = \sqrt{1 + p^2}$. Therefore, in this Breit frame the momentum transfer is

$$(p, E) - (-p, E) = (2p, 0), \tag{A.96}$$

with a zero energy component.

We can then write the proton form factor as

$$F(p) = \int e^{2ipz} \big(\psi_\eta(z, t)\big)^* \psi_{-\eta}(z, t) \ dz \, dt. \qquad (A.97)$$

Using the ground-state harmonic oscillator wave function, this integral becomes

$$\frac{1}{\pi} \int e^{2ipz} e^{-\cosh(2\eta)(z^2 + t^2)} \ dz \, dt. \qquad (A.98)$$

The physics of $\cosh(2\eta)$ in this expression was explained in equation (A.75).

The exponential function in the Fourier integral of equation (A.98), does not depend on the t variable. We can therefore integrate over t, and equation (A.98) becomes

$$F(p) = \frac{1}{\sqrt{\pi \cosh(2\eta)}} \int e^{2ipz} e^{-z^2 \cosh(2\eta)} \ dz. \qquad (A.99)$$

By completing this integral, the proton form factor becomes

$$F(p) = \frac{1}{\cosh(2\eta)} \exp\left(\frac{-p^2}{\cosh(2\eta)}\right). \qquad (A.100)$$

Using the expression of $\cosh(2\eta)$ given in equation (A.75), this proton form factor becomes

$$F(p) = \frac{1}{1 + 2p^2} \exp\left(\frac{-p^2}{1 + 2p^2}\right). \qquad (A.101)$$

This function decreases as $1/p^2$ for large values of p.

It is illustrative to demonstrate the effect of the role of this Lorentz contraction in more detail. To accomplish this we perform the integral of equation (A.99) without the contraction factor $\cosh(2\eta)$. Therefore, the wave function $\psi_\eta(z, t)$ in the equation (A.97) is replaced by the Gaussian form $\psi_0(z, t)$ of equation (A.67). This non-squeezed wave function results in the Fourier integral

$$G(p) = \int e^{2ipz} (\psi_0(z, t))^* \psi_0(z, t) \ dz \, dt. \qquad (A.102)$$

After integration the result is

$$G(p) = \frac{1}{\sqrt{\pi}} e^{-p^2}. \qquad (A.103)$$

From the above we see that this leads to a Gaussian cut-off of the proton form factor. Note that this does not happen in the real world, and the calculation of $G(p)$ is for illustration only.

By returning to the Fourier integrals of equations (A.97) and (A.102) we see that the only difference is the $\cosh(2\eta)$ factor in equation (A.97). Since this factor is in the normalization constant and comes from the integration over the t variable it does not affect the Fourier integral.

It does, however, cause the Gaussian width to shrink by $1/\sqrt{2}\,p$ for large values of p. This makes the wavelength of the sinusoidal factor inversely proportional to the momentum $2p$. Hence, both the Gaussian width and the wavelength of the incoming signal shrink at the same rate of $1/p$ as p becomes large. Without this coherence, the cut-off is Gaussian as noted in equation (A.103). This effect of Lorentz coherence is illustrated in figure A.7.

However there is a gap between $F(p)$ of equation (A.101) and the experimental data. Before comparing them however, it is necessary to realize that there are three quarks inside the proton with two oscillator modes.

We can then suppose that one of the modes goes through the Lorentz coherence process discussed in this section. The other mode must then go through the contraction process given in equation (A.75). This produces a net effect of

$$F_3(p) = \left(\frac{1}{1 + 2p^2}\right)^2 \exp\left(\frac{-p^2}{1 + 2p^2}\right). \tag{A.104}$$

The desired dipole cut-off of $(1/p^2)^2$ is thus obtained.

The effect of the quark spin should be also addressed. There have been reports of deviations from the exact dipole cut-off. This has resulted in attempts to study the proton form factors based on the four-dimensional rotation group with an imaginary time coordinate. Many papers based on lattice QCD have also been written. Some references for these efforts are given in [46].

This section was limited to studying the role of Lorentz coherence in keeping the proton form factor from the steep Gaussian cut-off in the momentum transfer

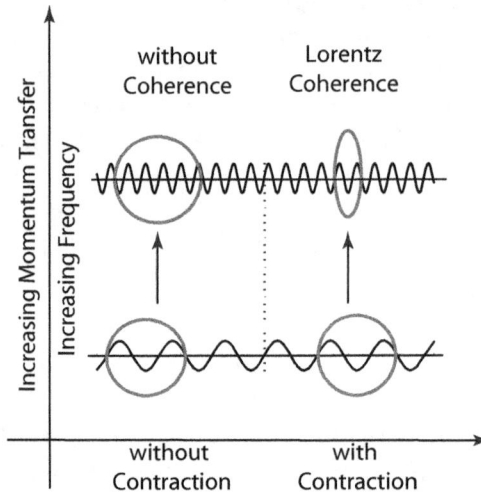

Figure A.7. Coherence between the wavelength and the proton size. Referring back to figure A.6, the proton sees the incoming photon. The wavelength of this photon becomes smaller for increasing momentum transfer. If the proton size remains unchanged, there is a rapid oscillation cut-off in the Fourier integral for the form factor leading to a Gaussian cut-off. However, if the proton size decreases coherently with the wavelength, there are no oscillation effects, leading to a polynomial decrease of the form factor [43, 46].

variable. One of the primary issues of the current trend in physics is this coherence problem.

A.2.4 Coherence in momentum–energy space

In order to study how Lorentz coherence manifests itself in momentum–energy space, we will start with the Lorentz-squeezed wave function in momentum–energy space. This can be written as

$$\phi_\eta(q_z, q_0) = \frac{1}{2\pi} \int e^{-i(q_z z - q_0 t)} \psi_\eta(z, t) dt\, dz, \tag{A.105}$$

where this is a Fourier transformation of the Lorentz-squeezed wave function of equation (A.67). Here q_z and q_0 are Fourier conjugate variables to z and t, respectively. The integration of the Fourier transform results in

$$\phi_\eta(q_z, q_0) = \frac{1}{\sqrt{\pi}} \exp\left\{-\frac{1}{4}\left[e^{-2\eta}(q_z + q_0)^2 + e^{2\eta}(q_z - q_0)^2\right]\right\}. \tag{A.106}$$

The proton form factor of equation (A.97) can now be written in terms of this momentum–energy wave function as

$$F(p) = \int \phi_{-\eta}^*(q_0, q_z - p)\phi_\eta(q_0, q_z + p) dq_0\, dq_z. \tag{A.107}$$

When this integral is evaluated the result leads to the proton form factor $F(p)$ given in equation (A.101).

To see the effect of the Lorentz coherence, consider the two wave functions in figure A.8. Since the integral is carried out over the $q_z q_0$ plane, when the momentum p increases, the two wave functions become separated. If it were not for the Lorentz-squeeze, the wave functions would not overlap, and this would lead to a sharp

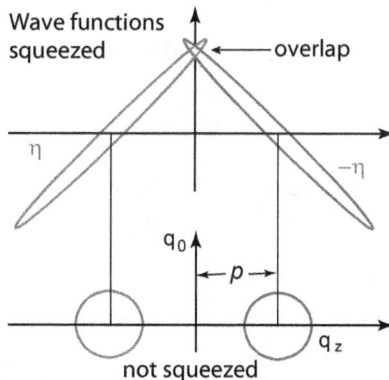

Figure A.8. Lorentz coherence in the momentum–energy space. Both squeezed and non-squeezed wave functions are given. As p increases, the two wave functions in equation (A.107) become separated. Without the squeeze, there are no overlaps. This leads to a Gaussian cut-off. The squeezed wave functions maintain an overlap, leading to a slower polynomial cut-off [43].

Gaussian cut-off as in the case of $G(p)$ of equation (A.103). However, the squeezed wave functions have an overlap as shown in figure A.8. This overlap becomes smaller as p increases and this leads to a slower polynomial cut-off [43, 46].

When the non-zero size of the proton was discovered [29, 30] it opened a new era of physics. As the proton is no longer a point particle, one way to measure its internal structure is to study the proton–electron scattering amplitude with one-photon exchange, and its dependence on the momentum transfer. The proton form factor discussed in the previous section measures this deviation from the case with the point-particle proton.

Experimentally, the dipole cut-off has been firmly established. There are also experimental results indicating deviations from this dipole behavior [2, 56]. Here no attempt has been made to review all the papers written on the corrections. Theoretically, those deviations are corrections from the basic dipole behavior.

The study of the proton form factor is still a major subject in physics. It is, however, gratifying to note that the proton's dipole cut-off comes from the coherence between the Lorentz contraction of the proton's longitudinal size and the decrease in the wavelength of the incoming signal.

A.2.5 Hadronic temperature

Almost all branches of physics find a way to use the harmonic oscillator wave functions. It is possible to excite the single-variable ground-state harmonic oscillator in the three following ways:

1. Energy level excitations, with the energy eigenvalues $\hbar\omega(n + 1/2)$.
2. Coherent state excitations resulting in

$$|\alpha\rangle = e^{-\alpha\alpha^*/2}e^{\alpha a^\dagger}|0\rangle = \sum_n \frac{\alpha^n}{\sqrt{n!}}|n\rangle. \tag{A.108}$$

3. Thermal excitations resulting in the density matrix of the form

$$\rho_T(z, z') = (1 - e^{-\hbar\omega/k_BT}) \sum_k e^{-k\hbar\omega/k_BT}\phi_k(z)\phi_k^*(z'), \tag{A.109}$$

where $\hbar\omega$ and k_B are the oscillator energy separation and Boltzmann's constant, respectively. This form of the density matrix is well known [9, 26, 39, 52].

In this section it is the thermal excitations which are of interest. We can write the density matrix of equation (A.109) as

$$\rho_T(z, z') = (1 - e^{-1/T}) \sum_k e^{-k/T}\phi_k(z)\phi_k^*(z'), \tag{A.110}$$

where the temperature is measured in units of $\hbar\omega/k_B$.

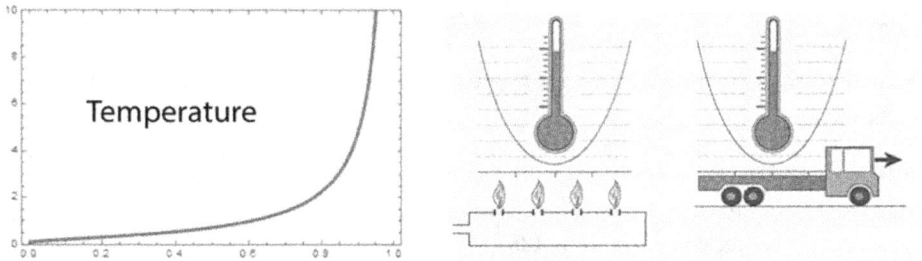

Figure A.9. Hadronic temperature plotted against β. As the hadron gains speed, the quarks inside become excited and this results in a rise in temperature. If the temperature is sufficiently high, those quarks start boiling and become partons [47].

Comparing this expression with the density matrix of equation (A.46), leads to

$$\tanh^2 \eta = e^{(-1/T)} \tag{A.111}$$

and to

$$T = \frac{-1}{\ln(\tanh^2 \eta)}. \tag{A.112}$$

We noted earlier in equation (A.47) that $\tanh(\eta)$ is proportional to the velocity of the hadron, and $\tanh(\eta) = v/c$. The oscillator thus is thermally excited as it moves. The calculation of the temperature as a function of $\tanh(\eta)$ together with the oscillator becoming thermally excited as it moves is plotted in figure A.9.

If we consider again the velocity dependence of the temperature, we see that it is almost proportional to the velocity from $\tanh(\eta) = 0$ to 0.7, and again from $\tanh(\eta) = 0.9$ to 1 with different slopes. The physical motivation for this section was based on Feynman's time separation variable [20] and his rest of the Universe [19]. It should also be noted that many authors have discussed field theoretic approaches to derive the density matrix of equation (A.109). Among those approaches are two-mode squeezed states of light [28, 44, 68, 70] and thermo-field-dynamics [16, 54, 57, 64].

Two-mode squeezed states share the same mathematics as that for the covariant harmonic oscillator formalism discussed in section A.1 [13, 28, 44, 70]. There are two measurable photons instead of the z- and t-coordinate. Should we decide not to observe one of them [15, 71], it belongs to Feynman's rest of the Universe [27].

It is also a remarkable feature of two-mode squeezed states of light that their formalism is identical to that of thermo-field-dynamics. The squeeze parameter in the two-mode case is related to the temperature. This is why it is possible to define the temperature of a Lorentz-squeezed hadron within the framework of the covariant harmonic oscillator model.

References

[1] Adesso G and Illuminati F 2007 Entanglement in continuous-variable systems: recent advances and current perspectives *J. Phys. A: Math. Theor.* **40** 7821–80

[2] Alkofer R, Höll A, Kloker M, Krassnigg A, and Roberts C D 2005 On nucleon electromagnetic form factors *Few-Body Syst.* **37** 1–31

[3] Başkal S, Kim Y S, and Noz M E 2005 *Physics of the Lorentz Group* (Bristol: IOP Publishing)

[4] Bég M A B, Lee B W, and Pais A 1964 $SU(6)$ and electromagnetic interactions *Phys. Rev. Lett.* **13** 514–7

[5] Bjorken J D and Paschos E A 1969 Inelastic electron–proton and γ–proton scattering and the structure of the nucleon *Phys. Rev.* **185** 1975–82

[6] Braunstein S L and van Loock P 2005 Quantum information with continuous variables *Rev. Mod. Phys.* **77** 513–77

[7] Buras A J 1980 Asymptotic freedom in deep inelastic processes in the leading order and beyond *Rev. Mod. Phys.* **52** 199–276

[8] Chou C-H, Yu T, and Hu B L 2008 Exact master equation and quantum decoherence of two coupled harmonic oscillators in a general environment *Phys. Rev.* E **77** 011112

[9] Davies R W and Davies K T R 1975 On the Wigner distribution function for an oscillator *Ann. Phys.* **89** 261–73

[10] Dirac P A M 1943 Quantum electrodynamics *Commun. Dublin Inst. Adv. Stud.* A **1** 36

[11] Dirac P A M 1945 Unitary representations of the Lorentz group *Proc. R. Soc.* A **183** 284–95

[12] Dirac P A M 1949 Forms of relativistic dynamics *Rev. Mod. Phys.* **21** 392–9

[13] Dirac P A M 1963 A Remarkable representation of the 3 + 2 de Sitter group *J. Math. Phys.* **4** 901–9

[14] Dodd P J and Halliwell J J 2004 Disentanglement and decoherence by open system dynamics *Phys. Rev.* A **69** 052105

[15] Ekert A K and Knight P L 1989 Correlations and squeezing of two-mode oscillations *Am. J. Phys.* **57** 692–7

[16] Fetter A L and Walecka J D 2003 *Quantum Theory of Many-Particle Systems* (Mineola, NY: Dover)

[17] Feynman R P 1969 The behavior of hadron collisions at extreme energies *Proc. 3rd Int. Conf. on High Energy Coll.* (*Stony Brook, NY* 5–6 September)Yang C N (New York: Gordon and Breach) pp 237–9

[18] Feynman R P 1969 Very high-energy collisions of hadrons *Phys. Rev. Lett.* **23** 1415–7

[19] Feynman R P 1998 *Statistical Mechanics: A Set of Lectures* (*Advanced Book Classics*) (Boulder, CO: Westview Press)

[20] Feynman R P, Kislinger M, and Ravndal F 1971 Current matrix elements from a relativistic quark model *Phys. Rev.* D **3** 2706–32

[21] Frazer W R and Fulco J R 1960 Effect of a pion–pion scattering resonance on nucleon structure. II *Phys. Rev.* **117** 1609–14

[22] Fujimura K, Kobayashi T, and Namiki M 1970 Nucleon electromagnetic form factors at high momentum transfers in an extended particle model based on the quark model *Prog. Theor. Phys.* **43** 73–9

[23] Gell-Mann M 1964 A schematic model of baryons and mesons *Phys. Lett.* **8** 214–5

[24] Giedke G, Wolf M M, Krüger O, Werner R F, and Cirac J I 2003 Entanglement of formation for symmetric Gaussian states *Phys. Rev. Lett.* **91** 107901-1–4

[25] Ginzburg V L, and Man'ko V I 1965 Relativistic oscillator models of elementary particles *Nucl. Phys.* **74** 577–88

[26] Han D, Kim Y S, and Noz M E 1990 Lorentz-squeezed hadrons and hadronic temperature *Phys. Lett.* A **144** 111–5

[27] Han D, Kim Y S, and Noz M E 1999 Illustrative example of Feynman's rest of the Universe *Am. J. Phys.* **67** 61–6

[28] Han D, Kim Y S, Noz M E, and Yeh L 1993 Symmetries of two-mode squeezed states *J. Math. Phys.* **34** 5493–508

[29] Hofstadter R 1956 Electron scattering and nuclear structure *Rev. Mod. Phys.* **28** 214–54

[30] Hofstadter R and McAllister R W 1955 Electron scattering from the proton *Phys. Rev.* **98** 217–8

[31] Hussar P E 1981 Valons and harmonic oscillators *Phys. Rev.* D **23** 2781–3

[32] Hussar P E, Kim Y S, and Noz M E 1985 Time–energy uncertainty relation and Lorentz covariance *Am. J. Phys.* **53** 142–7

[33] Hwa R C 1980 Evidence for valence–quark clusters in nucleon structure functions *Phys. Rev.* D **22** 759–64

[34] Hwa R C and Zahir M S 1981 Parton and valon distributions in the nucleon *Phys. Rev.* D **23** 2539–53

[35] Ishida S 1971 'Ur-citon': an attempt for unified theory of hadrons *Prog. Theor. Phys.* **46** 1905–23

[36] Kim Y S 1989 Observable gauge transformations in the parton picture *Phys. Rev. Lett.* **63** 348–51

[37] Kim Y S 1998 Does Lorentz boost destroy coherence? *Fortschr. Phys.* **46** 713–23

[38] Kim Y S 2004 Einstein, Wigner, and Feynman: from $E = mc^2$ to Feynman's decoherence via Wigner's little groups *Acta Phys. Hungarica* A **19** 317–28

[39] Kim Y S and Li M 1989 Squeezed states and thermally excited states in the Wigner phase-space picture of quantum mechanics *Phys. Lett.* A **139** 445–8

[40] Kim Y S and Noz M E 1973 Covariant harmonic oscillators and the quark model *Phys. Rev.* D **8** 3521–7

[41] Kim Y S and Noz M E 1975 Covariant harmonic oscillators and excited meson decays *Phys. Rev.* D **12** 129–38

[42] Kim Y S and Noz M E 1978 Quarks, partons, and Lorentz-deformed hadrons *Prog. Theor. Phys.* **60** 801–16

[43] Kim Y S and Noz M E 1986 *Theory and Applications of the Poincaré Group* (Dordrecht: Springer)

[44] Kim Y S and Noz M E 1991 *Phase Space Picture of Quantum Mechanics: Group Theoretical Approach* (*Lecture Notes in Physics* vol 40) (Singapore: World Scientific)

[45] Kim Y S and Noz M E 2005 Coupled oscillators, entangled oscillators, and Lorentz-covariant harmonic oscillators *J. Opt.* B **7** S458–67

[46] Kim Y S and Noz M E 2011 Lorentz harmonics, squeeze harmonics, and their physical applications *Symmetry* **3** 16–36

[47] Kim Y S and Noz M E 2014 Entropy and temperature from entangled space and time *Phys. Sci. Int. J.* **4** 1015–39

[48] Kim Y S, Noz M E, and Oh S H 1979 Lorentz deformation and the jet phenomenon *Found. Phys.* **9** 947–54

[49] Kim Y S, Noz M E, and Oh S H 1979 Representations of the Poincaré group for relativistic extended hadrons *J. Math. Phys.* **20** 1341–4

[50] Kim Y S, Noz M E, and Oh S H 1979 A simple method for illustrating the difference between the homogeneous and inhomogeneous Lorentz groups *Am. J. Phys.* **47** 892–7

[51] Kim Y S and Noz M E 1977 Covariant harmonic oscillators and the parton picture *Phys. Rev.* D **15** 335–8

[52] Landau L D and Lifshitz E M 2008 *Statistical Physics, Part 1. Course of Theoretical Physics* vol 5 3rd edn (Amsterdam: Elsevier)

[53] Licht A L and Pagnamenta A 1970 Wave functions and form factors for relativistic composite particles. I *Phys. Rev.* D **2** 1150–6

[54] Mann A and Revzen M 1989 Thermal coherent states *Phys. Lett.* A **134** 273–5

[55] Markov M 1956 On dynamically deformable form factors in the theory of elementary particles *Nuovo Cimento* **3** 760–72

[56] Matevosyan H H, Thomas A W, and Miller G A 2005 Study of lattice QCD form factors using the extended Gari–Krümpelmann model *Phys. Rev.* C **72** 065204-1–5

[57] Ojima I 1981 Gauge fields at finite temperatures—'thermo field dynamics' and the KMS condition and their extension to gauge theories *Ann. Phys.* **137** 1–32

[58] Paz J P and Roncaglia A J 2008 Dynamics of the entanglement between two oscillators in the same environment *Phys. Rev. Lett.* **100** 220401

[59] Rotbart F C 1981 Complete orthogonality relations for the covariant harmonic oscillator *Phys. Rev.* D **23** 3078–80

[60] Ruiz M J 1974 Orthogonality relation for covariant harmonic-oscillator wave functions *Phys. Rev.* D **10** 4306–7

[61] Sogami I 1973 Reconstruction of non-local field theory. I: causal description *Prog. Theor. Phys.* **50** 1729–47

[62] Takabayasi T 1964 Oscillator model for particles underlying unitary symmetry *Nuovo Cimento* **33** 668–72

[63] Takabayasi T 1979 Relativistic mechanics of confined particles as extended model of hadrons: the bilocal case *Prog. Theor. Phys. Suppl.* **67** 1–68

[64] Umezawa H, Matsumoto H, and Tachiki M 1982 *Thermo Field Dynamics and Condensed States* (Amsterdam: Elsevier)

[65] Wigner E P 1932 On the quantum correction for thermodynamic equilibrium *Phys. Rev.* **40** 749–59

[66] Wigner E P 1939 On unitary representations of the inhomogeneous Lorentz group *Ann. Math.* **40** 149–204

[67] Xiang S-H, Song K-H, Wen W, and Shi Z-G 2011 Entanglement behaviors of two-mode squeezed thermal states in two different environments *Eur. Phys. J.* D **62** 289–96

[68] Yuen H P 1976 Two-photon coherent states of the radiation field *Phys. Rev.* A **13** 2226–43

[69] Yukawa H 1953 Structure and mass spectrum of elementary particles. I. General considerations *Phys. Rev.* **91** 415–6

[70] Yurke B, McCall S L, and Klauder J R 1986 $SU(2)$ and $SU(1,1)$ interferometers *Phys. Rev.* A **33** 4033–54

[71] Yurke B and Potasek M 1987 Obtainment of thermal noise from a pure quantum state *Phys. Rev.* A **36** 3464–6

Index

3 + 2 deSitter group, 1.4.3

A

ABCD matrix, 9, 9.1–9.1.2, 9.2, 9.2.1
angular momenta, internal, 3
angular momentum, 4
annihilation operator, 1.4, 1.4.2, 6
anti-Hermitian, 1.4.1, 2, 2.2, 2.5
area-preserving transformation, 5, 5.5
attenuation, 9

B

Baker–Campbell–Hausdorff formula,
    5.7, 6.2, 6.4
Baker–Campbell–Hausdorff relation,
    6.2, 6.4
Baker–Campbell–Hausdorff theorem,
    6.2
Bargmann decomposition, 2.6, 9.1.2,
    9.1.3, 9.2.1
bilinear combinations, 4.4
bilinear conformal representation, 10
bilinear spinors, 4.4.2
bilinear states, 4.4
bilinear transformation, 2.7
Bloch equations, optical, 1.3
Bohr, Niels, 3
Bohr–Einstein question, A.2.1

C

camera optics, 9, 9.2.2
canonical transformations, 5.1, 6.5, 6.6,
    7, 7.4, 7.5, 7.6, A
Casimir operators, A.1.2
c-number, 5, 5.1, 5.3, 6.1
c-number commutator, 6.4
c-number operator, 5.7
coherence, A.2.2–A.2.4
coherence problem, A.2.3
coherency matrix, 11
coherent and squeezed states, 6.2

coherent state, 6, 6.3, 6.4, 6.5
coherent states of light, 6.2
commutation relations, Heisenberg, 1,
    1.4, 1.4.2, 1.4.3, 6.4
continuity problem, 3
coupled harmonic oscillators, 8
covariance, A.1.3
covariant, A.1.2, A.1.3
covariant harmonic oscillator, A.1.3,
    A.2.1, A.2.3
covariant harmonic oscillator
    formalism, A, A.1.3, A.1.4, A.2.2,
    A.2.5
covariant harmonic oscillator model, A.2.5
covariant harmonic oscillator wave
    function, A
covariant harmonic oscillators, A.1.3
covariant manner, A
covariant wave function, A.1.2
creation operator, 1.4, 1.4.2, 6
cylindrical group, 3.1.2
cylindrical transformation, 3.1.2

D

damped harmonic oscillator, 3, 3.5
decoherence, 11.2, 11.3.2
decoherency angle, 11.3.2
degenerate, infinitely, A.1.2
density matrix, 1, 1.3–1.3.3, 5, 5.1, 5.6,
    5.7, 7.5, 8, 8.1, 8.2, 8.3, 8.5, 8.6,
    11.3.4, A.1.3, A.1.4, A.2.5
density matrix, single variable, 5.1
dipole behavior, A.2.4
dipole cut-off, A.2.3, A.2.4
Dirac equation, 4
Dirac matrices, 7.4
Dirac spinor, 4.1
Dirac, P A M, 1.4.3
Dirac's form of relativistic quantum
    mechanics, A.1, A.1.2
distribution function, 5.1, 5.2, 5.3, 6.5

triangular matrix, 9.1.1
two-dimensional harmonic oscillator, 7
two-mode coherent photon state, 7
two-mode coherent state, 1.4.3
two-mode interferometers, 7

U
uncertainty relation, c-number, 6.6
uncertainty relation, Dirac–Heisenberg, A
uncertainty relation, Heisenberg, 1.4,
    1.4.1, 4, 5, 5.3, 6,
    6.1
uncertainty relation, phase-number, 6.1,
    6.6
uncertainty relation, time-energy, 6.1

V
vacuum state, 6

W
wave packet spread, 5, 5.2, 5.3
Wigner decomposition, 9.2, 9.2.1
Wigner frames, 3.1
Wigner function, 5, 5.1, 5.2, 5.3, 5.4, 5.5,
    5.6, 5.7, 6.5, 7.5, 7.6, 8.6, A.1.4
Wigner matrices, 3.1, 3.1.3, 3.2, 3.3, 3.4
Wigner momentum, 3.1, 3.1.3
Wigner rotation, 10.1.2
Wigner vector, 3.1, 3.2, 3.3,
    3.4
Wigner, Eugene P, 3

www.ingramcontent.com/pod-product-compliance
Lightning Source LLC
Chambersburg PA
CBHW080525220326
41599CB00032B/6209